GROWTH, DISTRIBUTION AND UNEVEN DEVELOPMENT

Growth, distribution, and uneven development

AMITAVA KRISHNA DUTT

Department of Economics

University of Notre Dame

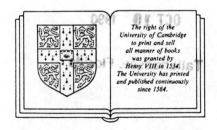

The right of the
University of Cambridge
to print and sell
all manner of books
was granted by
Henry VIII in 1534.
The University has printed
and published continuously
since 1584.

CAMBRIDGE UNIVERSITY PRESS

CAMBRIDGE

NEW YORK   PORT CHESTER

MELBOURNE SYDNEY

Published by the Press Syndicate of the University of Cambridge
The Pitt Building, Trumpington Street, Cambridge CB2 1RP
40 West 20th Street, New York, NY 10011, USA
10 Stamford Road, Oakleigh, Melbourne 3166, Australia

First published 1990

Printed at Interprint Ltd., Marsa, Malta

HD
75
D88
1990

British Library cataloguing in publication data
Dutt, Amitava Krishna
  Growth, distribution and uneven development.
  1. Economic development
  I. Title
  330.9

Library of Congress cataloguing in publication data
Dutt, Amitava Krishna.
  Growth, distribution, and uneven development/Amitava Krishna
Dutt.
    p.   cm.
  Includes index.
  ISBN 0-521-38177-0

  1. Economic development.   2. Resource allocation.
3. Technological innovations – Economic aspects.
HD75.D88   1990
338.9 – dc20   89-35683
                                                   CIP

ISBN 01 521 381770

IP

To my parents and my wife

CONTENTS

FIGURES

TABLES

ACKNOWLEDGEMENTS

Lance Taylor has shaped most of what I have written since the time I was formally his student, and the work presented here is no exception. While I owe him a incalculable intellectual debt, I am especially grateful here to him for his detailed and searching comments on earlier drafts. Geoffrey Harcourt has read and commented in detail on much of what I have written on the subject. He has been a constant source of encouragement and inspiration, without which this book would never have taken its present shape. Sandy Darity read the entire manuscript, and his challenging comments forced me to rethink many of my arguments. With Panos Liossatos I spent countless fruitful hours discussing many of the subjects dealt with in this book. Edward Amadeo and Paul Burkett made detailed and useful comments on several chapters.

I should also like to thank the CUP referees for their detailed reading of my manuscript. Lynn Mainwaring's suggestions and comments were particularly valuable.

Parts of the book were presented at seminars in several places, including the University of Calcutta, Presidency College (Calcutta), Indian Statistical Institute (Calcutta), the Massachusetts Institute of Technology, the Harvard-MIT joint seminar on non-neoclassical economics, the University of Denver, Florida International University, the Pontifical Catholic University at Rio de Janeiro, and the University of Notre Dame. I would like to thank their participants, as well as the discussants and other commentators at various conferences.

I should like to thank the students at Florida International University, at Pontifical Universidade Catolica de Rio de Janeiro, and at the University of Notre Dame. The suggestions made by Keun Lee and Ernesto Livacich were especially helpful. I am also grateful to Manjula Maudgal for her help in the later stages, and in preparing the index.

A summer research grant from the Provost's office at Florida International University made the writing of this book an easier task. Institutional support from Florida International University and the University of Notre Dame is also gratefully acknowledged.

This book incorporates some of the material published earlier in *Cambridge Journal of Economics, Eastern Economic Journal, Journal of Development Economics, Metroeconomica, Review of Radical Political Economics, Thames*

Papers in Political Economy, and *Zeitschrift fur Nationalokonomie.* I am grateful to their editors and anonymous referees for their comments.

I am grateful to Patrick McCartan and his editorial staff at CUP, especially Anne Rix, who did an excellent job of seeing this book to its conclusion.

Finally, I would like to thank my families in Calcutta and Miami for their unfailing interest and encouragement. I thank my wife for her constant support and required doses of interruptions and help.

CHAPTER 1

Introduction

1.1 Purpose of the book

The study of economic growth and income distribution has concerned many economists. It was central to the interests of the so-called classical economists, including Smith, Ricardo, and Marx. While the marginalist revolution turned the spotlight away to issues of resource allocation, its study went on with Schumpeter and the followers of Marx. More recently, the economics of growth has received a boost from the rise of Keynesian macroeconomics and the economics of less developed countries.

This book is concerned with the analysis of these issues of economic growth and distribution in capitalist economies, which we define as economies in which production, using hired labour, is privately organized with the aim of making profits.

The emphasis is on the analysis of growth, defined simply as an increase in total production (of goods and services). Our concern with distribution here is only to the extent that it relates to growth. We examine the issue of distribution, distinguishing between capitalists who own the material means of production and workers who work, by exploring how income distribution between these two classes is determined, and how it interacts with growth.

In addition to issues relating to growth and distribution *within* capitalist economies, this book is concerned with these issues in so far as they relate to relations *between* capitalist economies. It is thus concerned with growth at a world level, and with the possibility of the unequal distribution of its fruits between rich and poor countries. The discussion of these issues justifies the inclusion of the term 'uneven development' in the title of this book.

Hardly any new theories of growth or distribution are offered here, the emphasis being rather on taking stock of existing theories (some of which are more widely known than others). However, this is not done by surveying the literature. First, alternative approaches are viewed here from a particular methodological position which favours – in a sense – the demolition of walls separating economists partial to different approaches, by studying a general framework in which alternative approaches are interpreted as special cases. Secondly, the analysis is conducted with the explicit use of mathematical models, so that we occasionally examine new formalizations of existing

1

theories, sometimes clearing up cobwebs and extracting new implications. Thirdly, we are not concerned with the details of each approach but with what we view as their essence.[1] The first two observations suggest that this book intends to be more ambitious than a survey of existing theory; the last reveals that it is less comprehensive than one.

The method adopted here, following Sen (1963) and Marglin (1984a, 1984b), is to examine formally general frameworks which do not completely determine the behaviour of the economies they represent, because they result in underdetermined 'models', and then use alternative assumptions or closing rules to 'close' them to obtain alternative models. Most of this book is concerned with the examination of four such models, the neoclassical, neo-Marxian, neo-Keynesian, and Kalecki-Steindl models, no conscious attempt being made to favour any one, or to choose between them on empirical (or other) grounds.

It should be emphasized that no claim is being made that the theories examined here are in essence special cases of a more general approach, or are even best studied as if they are. They were certainly developed following different visions, and for different purposes. It is merely being claimed that examining them as alternative closures of a more general framework allows the construction of simple models representing each tradition, which sheds greater light on each approach, and allows simple comparisons to be made between them to the extent that they deal with comparable issues. Adherents of some of the traditions may argue that the models used in this book do not formalize their approach correctly. But rather than implying that the alternative-closures framework is misguided, this only suggests that the particular general framework and closing rule used may be inappropriate, in which case revisions may be made.[2]

1.2 Some methodological issues

Since the method followed here would seem to be at odds with the usual methodology adopted in economic analysis – developing theories within a particular research programme and subjecting them to empirical testing – a brief discussion of methodology is in order.

Popperian ideas are widely entertained by economists and influence most authors of economic principles texts.[3] The basic idea is that science progresses by generating theories which have excess empirical content, and by the process of empirical falsification; theories which are proved wrong ought to be discarded and new theories created. Our approach is not in keeping with this view of science, since we examine a variety of conflicting theories and make no effort to test them empirically.

Our reasons for parting company with the Popperian method do not require much belabouring, since the problems with it are well known. First, there are deep methodological problems behind the entire enterprise of hypothesis testing used in the statistical verification of theories.[4] Secondly, strict adherence to the falsification procedure would necessitate the jettisoning of all theories, and there is no agreement on how lenient the tests ought to be made: with some degree of leniency, and with some imagination exercised on the part of the econometrician, probably most theories would fail to be falsified. Thirdly, since all testing necessitates joint testing of a hypothesis with auxiliary hypotheses (the Duhem-Quine thesis) it is not possible to empirically falsify or corroborate any particular hypothesis.[5] Fourthly, and perhaps most important, testing, and theorizing are intrinsically related because empirical work requires concepts and categories which are themselves the results of particular theoretical perspectives. Often, as Feyerabend (1975) argues, the theories which are being tested themselves generate the definitions of theoretical concepts being measured, and this biases testing in their favour and against alternative theories. To make the falsification process work better we should give alternative theories some breathing space in which to develop, so that they could be in a position to generate their definitions. Given the dominance of neoclassical theory, definitions (for example, that of the concept of unemployment) used in empirical work are usually derived from it, biasing such work against alternative theories at this time.

We conclude that (at least for the present) Popperian ideas should not rigidly be applied in appraising different economic theories. In any case, more methodologically inclined economists subscribe instead to Kuhn's or Lakatos's approaches which use, respectively, the concepts of paradigms and research programmes. While Kuhn was interested primarily in understanding how science actually progressed and not in making prescriptions, it is possible to construct a weak Popperian approach, which would recommend retaining a paradigm – and continuing with normal science – for a while with the accumulation of falsified theories, until a scientific revolution or paradigmatic shift became necessary. This has all the problems that confound the Popperian method. But there are further problems with the concept of paradigms, which are not separately examined here, since they apply to Lakatos's concept of research programmes, to which we now turn.

Lakatos's approach views scientists as working in particular research programmes which have a hard core and a protective belt. The hard core contains propositions which are not even in principle testable. From these propositions are generated theories in the protective belt, which can be falsified, but falsification does not necessitate the rejection of what is in the hard core. Progress in a research programme takes place with the expansion of issues with which the programme deals. If we provisionally call the neoclassical, neo-Marxian, neo-Keynesian, and Kalecki-Steindlian approaches alternative research programmes (or paradigms), it seems that this book is not following Lakatosian (or Kuhnian) method, since we are not working within any one of them, but treating them as legitimate alternatives.

Our rejection of this method is related to the meaning of the hard core of a research programme (or the paradigm). It is not easy to find clear descriptions of what exactly are necessary and sufficient conditions for an economist or a theory to belong to a research programme.[6] The neoclassical programme has been most commonly studied,[7] and it seems that the essence of the programme is to use as an 'organizing principle' the postulate of optimizing behaviour by individual economic agents.[8] Thus 'explanation' in neoclassical economics consists in reducing what is to be explained to a model in which optimizing economic agents are examined, with full specification of their constraints and optimand. In line with this type of characterization of the neoclassical research programme, we can try to characterize other research programmes. While the neo-Marxian organizing principle can be said with some justification to be the use of 'class conflict' as a basis of explanation, it seems too difficult to go very far in finding clearcut organizing principles (of the same status as the optimizing principle) for the neo-Keynesian or Kalecki-Steindl research programmes.[9] We are thus led to one of two conclusions: the last two are not separate paradigms or research programmes, or our method of finding 'organizing principles' to define research programmes is ill conceived. On the first alternative, whether we call them different paradigms, research programmes or something else, this book will show that it is useful – and indeed vitally important – to treat the four as distinct alternatives. On the second, there can be no question that economists in different research programmes have placed a great deal of emphasis on their organizing principles.

While not disputing the fact that organizing principles have played important roles in the development of different types of theories, we would argue that they have created more disagreement than they should. To do so, we distinguish between 'organizing principles' and 'views of the economy' subscribed to by economists.[10] Assume that two economists – say a neoclassical and a neo-Marxist – are explaining the same set of events and considering their likely consequences. If they only differ in their organizing principle there is no reason why their explanations or forecasts should differ in content (apart from language). If they do differ, it is because they also hold different 'views of the economy', for instance, whether or not workers are (usually) fully employed (in some sense). It thus seems more sensible to distinguish between economists in terms of their views of the economy, and not their organizing principles.

But 'view of the economy' is a vague statement. We will give it some precision by relating it to the 'analytical dichotomy' the economist adopts, a dichotomy which is reflected in the decision to take certain things as 'givens' for the purposes of analysis. In formal terms, the economist may write down a large set of equations involving a large number of 'unknowns'. For the purposes of the theory, the economist adopts a particular dichotomy by deciding on which of the many equations are to be taken into consideration and which of the unknowns are to be taken as variables, the remaining unknowns being treated as data and held constant. The chosen equations then solve for the variables. The unknowns treated as data need not actually be fixed in reality: they may be treated as given in the particular theoretical work if one does not know – or does not think to be important – the relations between the data or the feedback effects from the variables to the data.[11]

The 'view of the economy' and 'analytical dichotomy' are related because the adoption of a particular dichotomy implies some particular vision of the economy, and conversely.[12] For instance, a dichotomy which takes the real wage of workers to be given usually implies the view that the economy has unemployed labour; one that takes as given the psychological motives of investors usually corresponds to the view that business psychology is extremely difficult to understand in terms of usual economic factors; and one taking as given the degree of monopoly power in an economy corresponds to the view that atomistic competition does not prevail in the economy.

It may be argued that 'organizing principles' and 'analytical dichotomies' which reflect 'views of the economy' are not independent. While we do not doubt that particular organizing principles have been associated with particular dichotomies and indeed, the adoption of a particular organizing principle may appear to make it more 'obvious' to adopt a particular dichotomy, we would insist on the difference. For instance, it may be tempting to go from the assumption that individuals maximize utility subject to constraints to the assumption that preferences are exogenous; but, while the first is the neoclassical organizing principle, the second is the result of a dichotomy – which takes tastes to be given – not logically implied by the former.

We have so far argued that research programmes (or paradigms) are best distinguished by the visions of the economy implicit in them, and that these visions are reflected in particular analytical dichotomies. We now argue that it is dangerous to work within the confines of one's research programme. While it is true that economists within a research programme adopt a particular dichotomy by taking certain things as given, some of the things taken as given can, at a later stage and when the need arises, be endogenized. Neoclassical economics has done this for endowments (in going from the exchange economy to the production economy) and later, for technology and preferences. Thus, it may be argued, economists working in any research programme must ultimately arrive at the same 'truth', by endogenizing everything that ought to be endogenized, so that there is no problem with working within a research programme. The problem, however, is that the method of endogenizing a particular set of data is not unique. When taking a step which endogenizes more elements of the set of data, a particular research programme will usually rule out certain kinds of changes. This may partly be due to the existence of vested interests which block new theories which seriously disturb the structure of older ones which treat new variables as data. But more subtly and commonly, it is the result of the rejection of extensions which have implications contrary to the 'intuitions' based on earlier theories. Thus, it may be argued that the neoclassical production model (which was appended to the neoclassical exchange model) was destined to be developed in a way that made its conclusions roughly similar to what was found in the exchange model (Pasinetti, 1981), making it irrelevant for the study of actual production economies. The same type of criticism (Davidson, 1978, Chick, 1983) has been made of models of monetary economies which originated from the neoclassical barter economy models. Such blocked extensions are clearly inimical to the proper development of a discipline, and biased 'progress' is inevitable in a discipline with only one research programme; the creativity and wisdom of a particular practitioner is also restricted. This implies that a discipline, and economics in particular, would gain not only by the simultaneous existence of alternative research programmes, but also from individual economists being aware of, and participating in, the development of alternative programmes. This is our motivation for not restricting ourselves to one research programme, and for examining alternatives.

This broadening of perspectives, which focuses on greater depth of explanation using insights from different research programmes, seems more important than what Lakatos meant by progress (a descriptive word with a prescriptive implication), that is, the ability to predict 'some novel, hitherto unexplained fact' (Lakatos, 1978), or to empirically verify it.[13] Indeed, some developments of the later type are no more than trivial theorizing of the type Kuhn took to be symptoms of a degenerating paradigm.

Rejecting the approaches of Popper and Lakatos (and Kuhn in a prescriptive interpretation), we are much closer to the approach of Feyerabend (1975): anything goes. Philosophers of science are incorrect in prescribing rules. To attain the fullest and best developments, individual creativity should not be restricted by rules imposed by non-practitioners. Feyerabend argued with historical illustrations (mainly from astronomy) that science progressed precisely because it did not follow the rules of what philosophers of science would call 'good method'. We have considered specific reasons why following established method would be destructive for the development of economics. We cannot agree with Blaug (1980) who writes that '[i]n the end, Feyerabend's book amounts to replacing the philosophy of science by the philosophy of flower power'. What it merely implies is that one must be aware of the methodological limitations of established methodological canons, and not to be restricted by them in one's research. To paraphrase Feyerabend, we could be neoclassicals one day, neo-Marxists the next, and follow Kalecki some other time.

The only problem, which Feyerabend does not address, is how one person can hope to be well versed in different types of theory. Our answer would be that it is better to specialize in areas or problems, rather than in research programmes. In this book we choose the area of growth and distribution, and examine work done on it in alternative research programmes.

We end this discussion with several comments about the method adopted in this book.

First, we shall examine alternative approaches to sharpen our insights on the determinants of growth and distribution and their interrelation, rather than insisting on choosing between them and finding one model to suit all situations. This approach is consistent with Solow's (1985) dictum:

> the true functions of analytical economics are best described informally: to organize our necessarily incomplete perceptions about the economy, to see connections that the untutored eyes would miss, to tell plausible – sometimes even convincing – causal stories with the help of a few central principles . . . In this scheme of things, the end product is likely to be a collection of models contingent on society's circumstances – on the historical context, you may say – and not a single monolithic model for all seasons.

Secondly, the approach followed is that of mathematical modelling. Shackle (1984) has distinguished between the axiomatic style 'which deals directly with an abstract system whose elements are defined only in relation to each other, so that it can only make suggestions about the world of experience by seeing in that world an analogy to itself in some respects', and the rhetorical style. He proceeds:

> The world of experience is unthinkably too various, rich, vague, complex and imprecise for logical demonstration to be possible by direct application to that world itself . . . The rhetorician, in common with the poet, is willing to use the full compass of meaning in any word he employs . . . The rhetorician deals, not in proof but in suggestion. Imprecision is, in his discourse, not a flaw, not a shortcoming, but an essential characteristic resource. It is the rhetorician's art to touch the harp-strings of individual imagination, to gain, not grammatical assent but a response, partly aesthetic, to find empathy in his reader's or his hearer's mind.

We have no disagreement with the need for rhetorical style, and we believe that it is useful in the creation of new ideas. But, given some ideas, it would be counterproductive not to attempt to make them precise by formalizing them. Not only could logical inconsistencies creep more easily into the rhetorical style than into the axiomatic one, but no means would be available to compare the essences of different approaches stripped of differences due to language. Translation into a common language, that of algebra, thus seems desirable when we wish to understand and compare alternative – already existing – theories of growth and distribution.[14]

Thirdly, though our approach eschews the use of organizing principles comparable to those in neoclassical or Marxian approaches, we use the type of modelling strategy – close to the work of Keynes, the Cambridge school, and the recently emergent mathematical structuralist school – that starts from definitional identities and then fills in the skeleton with macroeconomic behavioural relations. Like Keynes, we use macroeconomic aggregates and macroeconomic behavioural equations (such as the consumption function) not explicitly based on optimizing behaviour. Like the Cambridge economists we do use accounting relations between observables, but unlike them we use behavioural rules more explicitly to 'close' our models.[15] Like the structuralists we model specific features of the economy and give them a chance to affect the macroeconomic outcome, but unlike them we do not necessarily commit ourselves to any particular set of structural characteristics in examining a particular economy in a specific historical period.[16]

In not using more specific organizing principles, we do not mean to deny their usefulness, or claim that their use is necessarily problematic. As we shall see, many of the behavioural equations assumed in our models – given sufficient ingenuity – can be derived from optimizing behaviour, without changing the implications of the macroeconomic models in which they are embedded. Many of the components of the models can also be explained in terms of (that is, be reduced to) class struggle. Further, for many kinds of analysis, explicit optimization (in the study of many types of individual behaviour) or class struggle (in the study of broad historical patterns) are particularly useful and efficient. In our analysis we have not used them partly to suggest that in what we are concerned with, such principles are unnecessary. Also, using them unnecessarily restricts the analysis into doing certain things, and thus, given our bounded rationality, to the exclusion of other (in our judgement) more important matters.[17]

Finally, if for the progress of economics, we believe that everything is acceptable, we need to further justify the existence of this section on methodology, and indeed the book itself. Methodological anarchism does not imply methodological inactivism on the part of individuals, nor even *laissez faire*. The economics profession today is dominated by neoclassical economists who for whatever reasons – be they ideological or methodologically moral (against the informal approach adopted by many non-neoclassical economists)[18] – have chosen to exclude from the mainstream economists of other persuasions, not because they have shown them to be inferior, but because they do not follow their method. Since adherence to any single approach has been argued to be inimical to progress, sole adherence to the neoclassical approach is also destructive for economics. In this book, therefore, we have given at least equal billing to the other approaches, and usually a smaller role to the neoclassical one.

1.3 Equilibrium and causality

In examining growth, distribution and uneven development, this book will consider alternative *equilibrium* models, in the sense that they use a set of equations to solve for a set of unknowns (among which are the variables relating to growth and distribution) in terms of the forms of the equations and a set of data (which follow from a particular dichotomy). Despite the simplicity of this notion of equilibrium, given the many debates surrounding the concept, it is necessary to make some comments on it and the meaning of causality related to it.

First, this notion of equilibrium is completely abstract, and has no connection with any particular notion of equilibrium used in economic theory, such as the neoclassical one.[19]

Secondly, this notion is closely related to the one that stresses the balancing of forces. The same (equilibrium) outcome may result from different specifications of forces, but in some cases the forces will determine the properties of the resultant equilibria so that the study of forces becomes important. We will sometimes embark on this by examining the dynamics outside equilibria.

Thirdly, our conception of equilibrium does not take the economy to be actually at it: it could be hovering around it. Nevertheless we use the equilibrium notion to study actual economies because, if we did not, we would be able to say nothing about the economy.[20] Of course, if we could study the economy's 'hovering' in terms of some other set of data, some of which actually move over time according to systematic laws of motion, we would be considering another specific equilibrium where these laws of motion are ignored and their 'variables' are frozen. We shall often have occasion to distinguish between what standard economics calls short-run equilibrium and long-run equilibrium to study such cases: in the short run some of the variables of the long-run model are treated as data since they move slowly over time.[21] All this implies that, since different things change at different speeds, a continuum of different particular equilibria could be studied for the economy, making more things variables as we lengthen the 'run'. The notion of a particular 'long-run' or 'long-period position' as in classical political economy, is thus not sacrosanct. By endogenizing more elements of the data set we can study phenomena relating to change, which may themselves result in an equilibrium, or may end up with an unstable equilibrium, so that a succession of equilibrium positions of the economy (in terms of one set of data) does not converge to a final equilibrium (in terms of a smaller set of data). We shall examine the notion of uneven development between countries in these terms.

Fourthly, we do not assume that all the data are in reality constant. Some could be changing systematically through time, while others could be fluctuating haphazardly. There could also be some systematic relations between them. Any particular notion of equilibrium is thus a logical construction which chooses, for whatever reason, to hold some things as data and as not being explained within the model. The equilibrium therefore does not imply necessarily that the economy it is describing is in a tranquil state. Thus the individuals in the economy at equilibrium need not behave as if they knew the future.

Finally, the last two comments imply that we are treating time as logical or mechanical, rather than historical.[22] We cannot study the behaviour of actual economies, since they are too complex. We thus study movements which are notional, since they hold certain things as given, when we study the impact of changes in some of the data. For this kind of analysis to be meaningful, however, we should be careful in asking whether a particular change in data is likely to affect other data, a theme to which we will return later. When we have taken all the important and systematic influences into account we would be still talking about logical time – since we are dealing with abstract models and not history – but, hopefully, be taking into account Robinson's (1956, 1962) criticisms of neoclassical equilibrium theory.[23]

Given our notion of equilibrium, it is a simple matter to clarify what we mean by causation. We say that a particular rate of growth or income distribution is caused by the factors which explain the exogeneity of the elements of the set of data and equations. What causes these causes cannot be addressed by the particular model. By studying causality in this way we are not denying the existence of unidirectional causality flowing from some variables of our models to others. But such causality makes sense in some other type of model and consequent equilibrium notion which employs a different analytical dichotomy.[24]

1.4 Outline of the book

The rest of this book proceeds as follows.

Chapter 2 presents a general framework for examining a one-sector closed economy without inflation and technological change, which is then used to examine the long- and short-run behaviour of the four basic models with which the book is concerned, the neoclassical, the neo-Marxian, the neo-Keynesian, and the Kalecki-Steindl, in addition to others.

Chapters 3 through 5 modify the basic models in several ways. Chapter 3 relaxes some of the simplifying assumptions, mainly to justify the names given to them, while chapters 4 and 5, respectively, introduce money and inflation, and technological change into them to highlight some additional differences between the alternative approaches.

Chapters 6 and 7 extend the analysis to consider two sectors, producing a consumption good and an investment good. Chapter 6 develops two-sector versions of the basic models, as well as others, first assuming classical long-period equilibrium and then examining disequilibrium dynamics. The analysis of these models allows us to discuss, in chapter 7, a variety of doctrinal issues, including those relating to the nature of classical political economy, Sraffian analysis, and competition versus monopoly power

The next three chapters extend our analysis to consider trade between two economies – a rich North and a poor South – to explore mechanisms of uneven development. Chapter 8 introduces the general framework for the two-region world and develops alternative models, and chapters 9 and 10 examine specific models to show how endogenous preferences and technological change can result in uneven international development.

Chapter 11 concludes.

CHAPTER 2

Alternative models of growth and distribution

2.0 Introduction

This chapter examines the basic framework employed in the rest of the book for the analysis of growth and income distribution. The framework, for a closed economy producing only one good, is developed in section 2.1. The relations developed there do not determine the values of the relevant variables (which is why it is called a framework rather than a model); four alternative ways of 'closing' the framework will be explored in section 2.2, yielding the neoclassical, neo-Marxian, neo-Keynesian, and Kalecki-Steindl models. While these models are examined only in their long-run equilibrium states in this section, section 2.3 examines their behaviour in the short run and their dynamics over time.

2.1 The general framework

We make the following assumptions:

(A.1) The economy is closed.

(A.2) The economy produces only one good, which can be used for consumption or investment.

(A.3) The economy is capitalist in the sense that firms produce for profit with hired labour. Investment goods are held by the firms (since there is no rental market for it).

(A.4) Production takes place with *given* input-output relations. For want of better terminology, we shall refer to these relations as technology, although they may depend on factors other than purely technological ones.

(A.5) There are two factors – homogeneous labour and capital (physically the same as the produced good) – of production. There are no other factors such as land or intermediate goods.

(A.6) Technology exhibits fixed coefficients and constant returns to scale.

(A.7) There is no government fiscal activity.

(A.8) Monetary and asset market consideration are ignored.

(A.9) Capital does not depreciate.

(A.10) All firms are identical.

Assumptions (A.1) and (A.2) are dispensed with after chapter 5. Assumption (A.3) will be made throughout the book, except for some parts dealing specifically with non-capitalistic production sectors. The other assumptions are made for simplicity; the effects of modifying some of them are studied in various parts of the book to provide a flavour of what they imply. Assumption (A.10) allows us to use the concept of the representative firm and ignore the problem of aggregation.

Given these assumptions, we may examine the quantity and price sides of the economy by deriving a production equation and a price equation.[1]

Starting with the quantity side, output can either be consumed or invested, so that

$$X = CL + gK,$$

where X is output, L the amount of employment, K the amount of capital, C the consumption per employed worker, and g the rate of growth of capital. Dividing by X and denoting the employment-output ratio given by the fixed coefficients technology by a_0, we get

$$1 = Ca_0 + g(K/X) \tag{2.1}$$

which is our production equation. Note that the asymmetry which allows excess capital to exist, but the labour-output ratio to be fixed technologically, results from the fact that labour is hired (and will therefore not be hired if it does not contribute to production) and capital is not (so that excess capital may be held by firms).

If a_1 is the technologically fixed capital-output ratio we must have

$$K/X \geqslant a_1, \tag{2.2}$$

where the equality implies production at full capacity, and the strict inequality, excess capacity of capital. For the case of $K/X = a_1$ we get $1 = Ca_0 + ga_1$. If the economy operates at full capacity utilization it must be on the line representing this equation, which we call the potential consumption-growth frontier (see Figure 2.1); it has C and g intercepts $1/a_0$ and $1/a_1$, respectively, and a slope of $-(a_1/a_0)$. It follows from (2.2) that combinations of C and g must lie in $\langle C, g \rangle$ space in the closed set enclosed by this line and the axes. For given $K/X > a_1$, equation (2.1) can be represented by a line with C and g intercepts given by $1/a_0$ and X/K and slope $-(K/X)/a_0$; the lower is X/K, the steeper the curve, with a fixed C intercept. We will call them consumption-growth frontiers.

Turning now to the price side, price per unit either goes to wages or to profits, so that

$$P = Wa_0 + rP(K/X),$$

where P is price, W is the money wage, and r the rate of profit, defined as the value of profit divided by the value of capital, so that rPK is total profit. We assume that the profit rate is defined in terms of current price, and with wages paid after production, the wage fund is not included as capital which earns the

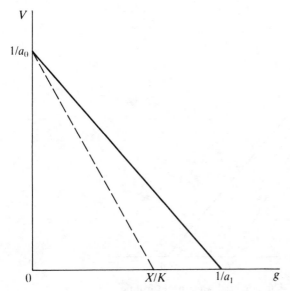

Figure 2.1   The potential consumption-growth frontier

profit. Dividing through by P we get

$$1 = Va_0 + r(K/X), \tag{2.3}$$

where $V (= W/P)$ is the real wage, which is our price equation. Remembering (2.2) we find that combinations of V and r must lie in the closed set in $\langle V, r \rangle$ space bounded by $1 = Va_0 + ra_1$ and the axes. This is shown in Figure 2.2. Analogously with the case of quantities, the line given by this equation will be called the potential wage-profit frontier, and the line given by (2.3) for any given K/X will be called the wage-profit frontier. The slopes and intercepts of these curves are exactly the same as those of the potential consumption-growth frontier and consumption-growth frontier, respectively.

The equations contain five variables, C, g, V, r, and K/X, the levels of which measure the rate of growth and distribution in the economy (although we need more information on who earns the wage and profit incomes to learn about personal income distribution), as well as consumption levels and use of productive capacity. We need to determine their values to understand the determination of growth and distribution (among other things) in this economy.

We define a *long-run equilibrium* as a set of values for these variables which are completely determined by a set of equations.[2] Our two equations (and inequality (2.2)) cannot determine this equilibrium; they tell us that the equilibrium values of C, g, V, and r must lie inside or on the potential consumption-growth and potential wage-profit frontiers but not exactly

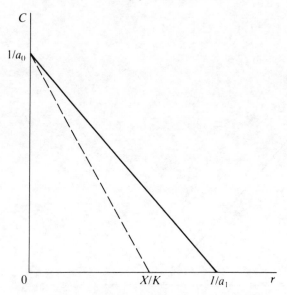

Figure 2.2   The potential wage-profit frontier

where. We need additional equations to solve for the long-run equilibrium.

The two equations actually incorporate very little *theoretical* content. Assumptions (A.1) through (A.10) do not specify the pricing, production, and investment policies of firms, what income recipients do with their income, and the nature of the goods and labour markets. In the absence of all this information it is not surprising that we cannot determine the equilibrium.

To make the framework more specific, we add two assumptions:

(A.11) There are two classes in the economy: capitalists and workers. Workers perform the labour required for production and capitalists do not (except as supervisors).

(A.12) Workers consume all their income, and capitalists save a constant fraction, *s*, of their income.

Assumption (A.11) implies that only workers can earn wage income; coupled with (A.12) it implies that capitalists earn the entire profit income. They provide the needed link between functional and personal income distribution and are intended to capture the stylized fact that capitalists, who earn profit income, are rich and have a higher propensity to save than workers, who are poor. We have gone to the extreme of assuming that workers save nothing, which enables us to ignore complications arising from the facts that workers may save, hold capital, and earn profits.[3]

These assumptions imply that

$$PCL = WL + (1-s)rPK$$

which, divided by PX and substituted into (2.1), implies

$$g = sr \qquad (2.4)$$

which states that investment equals saving (both as ratios of capital stock).

The system of equations (2.1), (2.3), and (2.4) will provide our general framework. Since we have five variables to solve for, we need two more equations to solve them. The next section will examine four alternative sets of two equations which can close our underdetermined 'model', which will be identified with the neoclassical, neo-Marxian, neo-Keynesian, and Kalecki-Steindl theories of growth and distribution.

We end with the remark that the general framework has already become rather specific. While some of the assumptions are rather innocuous and only adopted for the sake of simplicity, a few of them may already seem to be inconsistent with the visions underlying some of the theories mentioned in the preceding paragraph. Thus, the fixed coefficients assumption and the saving behaviour postulated may be thought to be extremely anti-neoclassical and the assumption of given technology may be thought to be anti-Marxian. Indeed, the whole method of determining long-run equilibrium may be thought to be contrary to the intentions of Marx (who was interested in analysing historical transformations encompassing the interplay of a large variety of factors) or the neo-Keynesians (who are interested in dynamic processes in history) or Kalecki (who was primarily interested in fluctuations). Some of the specific assumptions (about technology and consumption behaviour) are made to provide a simple common framework, and will be relaxed in the subsequent discussion of some of the models. The long-run framework is adopted to serve as a common feature of the models which enables the easy comparison of the theories on issues on which they all have something to say. Short-run features of the models will be considered later, and our analysis of the short run and the long run can be thought of as providing a starting point for analysis which can be extended to a wider canvas, some examples of which are suggested in the course of this book.

2.2 Alternative closures

This section examines four models resulting from alternative sets of two equations which can 'close' the underdetermined framework. Here we concentrate on their formal structures and postpone discussion of the approaches they represent, and justification of the names given to them, to the next chapter.

2.2.1 *The neoclassical closure*

The neoclassical closure additionally assumes:

(NC.1) Perfect competition prevails in the goods market.
(NC.2) There is no unemployment of labour in long-run equilibrium.

Assumption (NC.1) implies that all producing firms are atomistic price takers and that the price varies to clear the goods market. Assumption (NC.2) is at a different level since it is an assumption about a result. However, it may be taken to follow from the assumption of wage flexibility, which keeps the demand and supply of labour in balance in long-run equilibrium.[4]

We also make the following simplifying assumption:

(A.13) The supply of labour grows at an exogenously fixed rate, denoted by n.

Assumption (A.13) will be used, when necessary, in the other closures as well, and is not specific to the neoclassical model. It can be justified in terms of the constancy of the rate of population growth and the participation ratio (determined, respectively by demographic and sociological factors outside the scope of our analysis).

These additional assumptions provide us with two new equations which can 'close' the system. Assumption (NC.1) implies that firms will produce at full capacity since they believe that they can sell any amount they want to at the going price (that is, if they can obtain enough labour, which we shall assume for now); thus

$$K/X = a_1. \tag{2.5}$$

Assumption (NC.2) implies that in long-run equilibrium the supply of labour is equal to the demand for labour and that they must grow at the same rate; since by (A.13) supply grows at rate n, and by (A.6) and equation (2.5) the demand grows at the rate of capital accumulation, we have[5]

$$g = n. \tag{2.6}$$

Equations (2.1) and (2.3) through (2.6) are five equations which can solve for the five variables. Equation (2.5) determines $(K/X)^*$, (2.6) fixes g^*, substitution of (2.5) in (2.1) solves for C^*, (2.4) solves for r^*, and substitution of (2.5) in (2.3) then solves for $\dot{V}^*$ where asterisks denote equilibrium values. Figure 2.3 depicts this equilibrium, where sr denotes (2.4) and the upper quadrants show the wage-profit and consumption-growth possibility frontiers.

The equilibrium values of the five variables are given by[6]

$$(K/X)^* = a_1,$$
$$g^* = n,$$
$$r^* = n/s, \tag{2.7}$$
$$C^* = (1 - na_1)/n_0,$$
$$V^* = [1 - (n/s)a_1]/a_0.$$

The wage-profit rate ratio is given by

$$(V/r)^* = [(s/n) - a_1]/a_0,$$

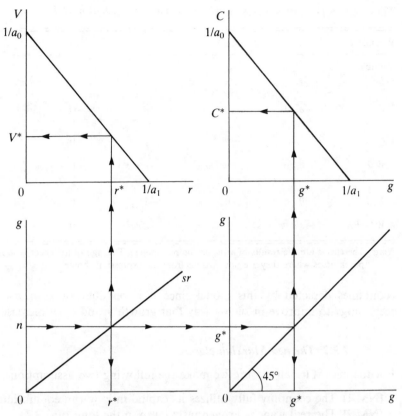

Figure 2.3 Long-run equilibrium in the neoclassical model

and that ratio of the wage bill to profit is given by

$$(WL/rPK)^* = (s/na_1) - 1.$$

A meaningful equilibrium requires positive values of all the variables. For $V^* > 0$ we require $n < s/a_1$, which also satisfies the weaker condition, $n < 1/a_1$, required for $C^* > 0$. Note that even with this condition satisfied there is no guarantee that the real wage will meet some minimum subsistence level; to ensure this we require $n \leqslant (1 - \tilde{V}a_0)s/a_1$, where $\tilde{V}$ denotes the subsistence wage.

The effects of parametric shifts are summarized in Table 2.1. The only parametric shift which can alter the rate of growth of the economy is n. Other things constant, a rise in n will increase g^* and raise r^* (to increase the rate of saving to make the faster rate of investment possible) and reduce V^* (to make the increase in r^* possible). We thus find an inverse relation between the real wage and the rate of profit, and an inverse relation between the rate of growth and the real wage, though this inverse relation need not hold over time in real

Table 2.1. *Effects of parametric shifts in the neoclassical model*

Parameter	n	s	a_1	a_0
Variable				
g	1	0	0	0
C	$-a_1/a_0$	0	$-n/a_0$	$-(1-na_1)/a_0^2$ $(-)$
r	$1/s$	$-n/s^2$	0	0
W/P	$-a_0/sa_1$	na_1/a_0s^2	$-n/a_0s$	$-[1-(n/s)a_1]/a_0^2$ $(-)$
X/K	0	0	$-1/a_1$	0
WL/rPK	$-s/a_1n^2$	$1/na_1$	$-s/na_1^2$	0

Note: The effects are the results of a rise in the parameters. The sign of the effect is shown in parentheses where they are not obvious from the expression shown.

economies depicted by this model since it is possible for n to rise and technology to improve in such a way that growth, r and V all increase.

2.2.2 *The neo-Marxian closure*

For the neo-Marxian closure we make the following two assumptions:[7]

(NM.1) The economy fully utilizes its capital in long-run equilibrium.
(NM.2) The real wage is exogenously given in the long run.

Assumption (NM.1) implies that (2.5) is satisfied, which is equivalent to the implication of assumption (NC.1). Assumption (NM.1) is stated differently from (NC.1) because neo-Marxist writers do not necessarily assume that firms are price takers,[8] although they do assume full utilization of capacity (perhaps due to efforts by firms of increasing market shares).

Assumption (NM.2) can be called a subsistence wage assumption, although the exogenously fixed level could depend on Marx's moral and historical elements, not just biology. Below it, workers will not work but join the ranks of the reserve army, living off some kind of activity outside the capitalist system. To make the model consistent it must be assumed that the demand for labour must not be large enough to exhaust the reserved army. If there is unemployment initially, and (A.13) is assumed, we require $g \leqslant n$. If this is violated, eventually the reserve army will be exhausted, there will be an upward pressure on the real wage and the economy may become neoclassical as in 2.2.1.[9] Assumption (NM.2) implies

$$V = \tilde{V}, \tag{2.8}$$

where $\tilde{V}$ is the level of the fixed real wage.

Equations (2.1), (2.3) to (2.5), and (2.8) are five equations which can solve for our five variables. Equation (2.5) solves for $(K/X)^*$, which can be substituted into (2.1) and (2.3); (2.8) solves for V^*; (2.3) consequently solves for r^*, which, when substituted into (2.4) solves for g^*; (2.1) finally solves for C^*. Figure 2.4 provides a diagrammatic depiction.

The equilibrium values of the variables (and the derived income distribution indicators) are given by

$$
\begin{aligned}
(K/X)^* &= a_1, \\
V^* &= \tilde{V}, \\
r^* &= (1 - \tilde{V}a_0)/a_1, \\
g^* &= s(1 - \tilde{V}a_0)/a_1, \\
C^* &= s\tilde{V} + (1 - s)/a_0, \\
(V/r)^* &= a_1/[(1/\tilde{V}) - a_0], \\
(WL/rPK)^* &= 1/[(1/\tilde{V}a_0) - 1].
\end{aligned}
\tag{2.9}
$$

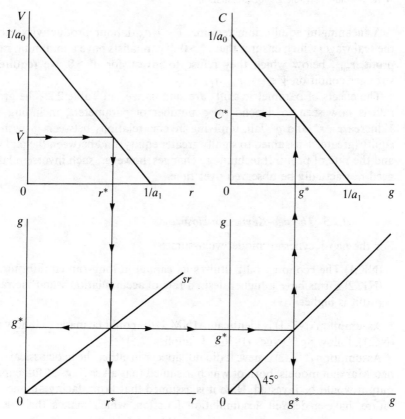

Figure 2.4   Long-run equilibrium in the neo-Marxian model

Table 2.2. *Effects of parametric shifts in the neo-Marxian model*

Parameter / Variable	$\tilde{V}$	s	a_1	a_0
g	$-sa_0/a_1$	$(1-\tilde{V}a_0)/a_1$ $(+)$	$-s(1-\tilde{V}a_0)/a_1^2$ $(-)$	$-\tilde{V}/a_1$
C	s	$\tilde{V}-(1/a_0)$ $(-)$	0	$-(1-s)/a_0^2$ $(-)$
r	$-a_0/a_1$	0	$-(1-\tilde{V}a_0)/a_1^2$ $(-)$	$-\tilde{V}/a_1$
W/P	1	0	0	0
X/K	0	0	$-1/a_1$	0
WL/rPK	$\dfrac{a_0}{[(1/\tilde{V}a_0)-1]^2(1/\tilde{V}a_0)^2}$	0	0	$\dfrac{a_0}{[(1/\tilde{V}a_0)-1]^2(1/\tilde{V}a_0)^2}$

Note: See note in Table 2.1 for explanation.

A meaningful equilibrium requires $\tilde{V} < 1/a_0$ (labour productivity exceeds the real wage) which ensures that $r^* > 0$. If capitalists have a minimum rate of profit, $r_{\min}$, below which they refuse to invest, for $g^* \geqslant 0$, we require the stronger condition $\tilde{V} \leqslant (1 - r_{\min} a_1)/a_0$.

The effects of parametric shifts are summarized in Table 2.2. The growth rate is now seen to depend on a number of parameters, including $\tilde{V}$. If $\tilde{V}$ increases, r^* and g^* fall, implying inverse relations between growth and equity (greater $\tilde{V}$ assumed to signify greater equity) and between the real wage and the rate of profit. If technology changes, however, such inverse relations need not actually be observed over time.

2.2.3 *The neo-Keynesian closure*

For the neo-Keynesian model we assume:

(NK.1) The economy fully utilizes its capital in long-run equilibrium.
(NK.2) Firms have a higher desired rate of accumulation when the rate of profit is higher.

Assumption (NK.1) is identical to (NM.1) (or could be made to follow from (NC.1), following Keynes (1936)). It implies (2.5).

Assumption (NK.2) is new; it did not appear in either the neoclassical or the neo-Marxian models, both of which assumed that all savings at full capacity output would be invested. Here it is assumed that firms, facing an uncertain future, have a desired accumulation function, which makes their desired investment depend positively on the expected profit rate. We further assume that the expected profit rate is equal to the actual one,[10] and that at equilibrium desired and actual investment are equal.[11] These yield (NK.2), which we write as

$$g = g(r) \quad g' > 0. \tag{2.10}$$

The position of this function depends on business psychology and on the historical and political characteristics of the economy, representing what Keynes (1936) and Robinson (1962) referred to as 'animal spirits'. We will assume that $g'' \leqslant 0$ and for simplicity, that $g(0) > 0$; implications of the relaxations of these assumptions will be considered later.

The neo-Keynesian model consists of equations (2.1), (2.3) through (2.5), and (2.10), which solve for the five variables. Equations (2.4) and (2.10) imply

$$g(r) = sr \tag{2.11}$$

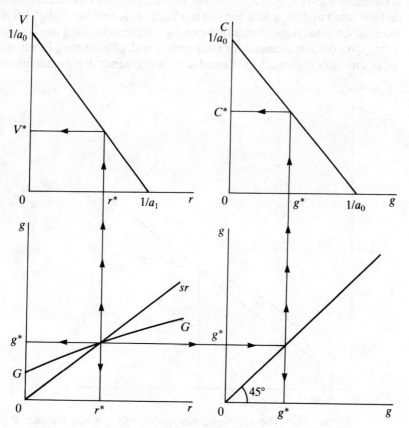

Figure 2.5 Long-run equilibrium in the neo-Keynesian model

which solves for r^* and g^*. Equation (2.5) can be substituted into (2.1) and (2.3) and substitution of g^* and r^* into them solve for C^* and V^*. Figure 2.5, where the GG curve depicts (2.10), shows the determination of equilibrium.

The equilibrium values of the variables are seen to be

$$(K/X)^* = a_1,$$
$$r^* \mid sr^* = g(r^*),$$
$$g^* = g(r^*),$$
$$C^* = (1 - g^* a_1)/a_0,$$
$$V^* = (1 - r^* a_1)/a_0,$$
$$(V/r)^* = [(1/r^*) - a_1]/a_0,$$
$$(WL/rPK)^* = (1/r^* a_1) - 1.$$

$$(2.12)$$

Given $g(0) > 0$ and $g'' \leqslant 0$, g and sr functions will intersect. To ensure $V > 0$, it is sufficient that they intersect at $r^* < 1/a_1$, but if V must be bounded below by a subsistence level $\tilde{V}$, then $r^* \leqslant (1 - \tilde{V}a_0)/a_1$ is required, which requires that s is not too small and the g function not too 'high'. It is obvious that $g'' \leqslant 0$ is not necessary; all that is required is that g and sr intersect satisfying this inequality. Also, $g(0) > 0$ is not necessary for intersection: with $g(0) < 0$, the g function may be as shown in Figure 2.6 (perhaps due to the existence of a minimum rate of

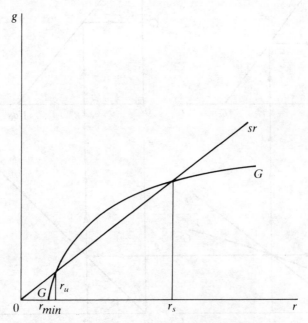

Figure 2.6   The neo-Keynesian model with a minimum rate of profit requirement

Table 2.3. *Effects of parametric shifts in the neo-Keynesian model*

Parameter	E	s	a_1	a_0
Variable				
g	$s/(s-g')$ $(+)$	$-g'g/s(s-g')$ $(-)$	0	0
C	$-[s/(s-g')]a_1/a_0$ $(-)$	$g'ga_1/s(s-g')a_0$ $(+)$	$-g/a_0$	$g/a_1a_0^2$
r	$1/(s-g')$ $(+)$	$-r/(s-g')$ $(-)$	0	0
W/P	$-a_1/(s-g')a_0$ $(-)$	$a_1r/(s-g')a_0$ $(+)$	$-r/a_0$	$r/a_1a_0^2$
X/K	0	0	$1/a_1$	0
WL/rPR	$-1/(s-g')a_1r^2$ $(-)$	$1/(s-g')a_1r$ $(+)$	$-1/ra_1^2$	0

Note: The desired accumulation function is written as $g = g(r, E)$ with $\partial g/\partial r = g'$ and $\partial g/\partial E = 1$. See also the note in Table 2.1.

profit, r_{min}, below which no accumulation will be desired), so that equilibrium will not be unique, but under reasonable adjustment patterns out of equilibrium (see the next section), the lower equilibrium will not be stable.[12]

The effects of parametric shifts are summarized in Table 2.3. The rate of growth is determined by the parameters of the g function and s. An upward shift in the g function will increase r^* and g^*, and reduce V^* and C^*: more potent animal spirits stimulate growth, but this is possible only due to forced saving resulting from a redistribution of income towards high-saving profit-earners. An increase in s reduces g^* and r^*, but raises V^* and C^*, so we have the paradox of thrift. These examples show that here growth and distribution, and the rate of profit and the real wage, are inversely related given the other parameters.

2.2.4 The Kalecki-Steindl closure

This closure assumes:[13]

(KS.1) That firms set the price level as a markup on prime costs; the markup rate is given.

(KS.2) Firms have a higher desired rate of accumulation if, *ceteris paribus*, the profit rate is higher, or, *ceteris paribus*, the rate of capacity utilization is higher

Assumption (KS.1) makes an explicit assumption that firms are not price-takers in a perfectly competitive economy. With fixed coefficients and constant returns to scale, this implies that firms set prices once W is given (which we assume for simplicity until chapter 4). This follows Kalecki (1971) in assuming that firms fix their price and adjust quantities in response to changes in demand. If demand is not enough to sell full capacity output, rather than change price they reduce output so that excess capacity results. The price becomes independent of demand, and is set by firms as a markup on prime costs, Wa_0, so that

$$P = Wa_0(1 + z), \tag{2.13}$$

where z, the rate of markup, is a parameter, given by the degree of monopoly power the level of which depends, among other things, on the extent of industrial concentration.

With (KS.1) paving the way for the emergence of excess capacity so that $X/K < 1/a_1$, it is natural to expect that, other things constant, a higher desired rate of accumulation is associated with a higher utilization of capacity.

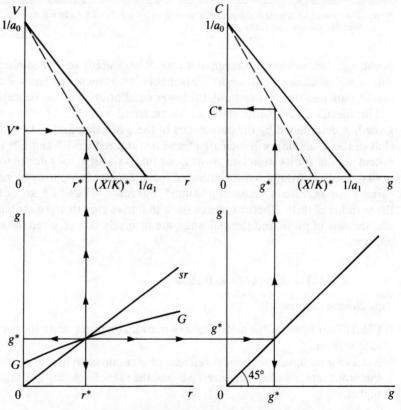

Figure 2.7   Long-run equilibrium in the Kalecki–Steindl model

Assumption (KS.2) follows Steindl (1952) in assuming this type of behaviour. It implies

$$g = g(r, X/K),\qquad(2.14)$$

where X/K denotes the degree of capacity utilization, and both partial derivatives are positive.[14]

Equations (2.1), (2.3), (2.4), (2.13), and (2.14) solve for the values of the five variables. Equation (2.13) solves V^* which, substituted into (2.3) gives

$$r = [z/(1+z)]X/K.\qquad(2.15)$$

Equations (2.4), (2.14), and (2.15) imply

$$g(r, [(1+z)/z]r) = sr\qquad(2.16)$$

which solves for r^* and g^*. Equation (2.15) then solves for $(X/K)^*$ and substitution of these values in (2.1) solves for C^*. All this is graphically depicted in Figure 2.7. Note that the GG curve in the third quadrant shows the left-hand side of (2.16), so that it is upward rising because a rise in r stimulates g both directly and indirectly through an implied increase in X/K. The intersection of this curve with sr determines g^* and r^*; this r^* and the value of V^* (from (2.13)) determine the position of the economy inside the wage-profit frontier of the first quadrant and the horizontal intercept of the dashed wage-profit frontier shows the degree of capacity utilization.

The equilibrium values of the variables are

$$
\begin{aligned}
V^* &= 1/a_0(1+z),\\
C^* &= [1+(1-s)z]/(1+z)a_0,\\
r^* &\mid g(r^*, [(1+z)/z]r^*) = sr^*,\\
g^* &= sr^*,\qquad(2.17)\\
(X/K)^* &= [(1+z)/z]r^*,\\
(V/r)^* &= 1/a_0(1+z)r^*,\\
(WL/rPK)^* &= 1/z.
\end{aligned}
$$

For a meaningful equilibrium we not only need curves GG and sr to intersect, but to intersect at an r such that, given the value of V, the V-r configuration lies *inside* the potential wage-profit frontier to allow excess capacity to exist. This requires us to impose some restrictions on the model. If (2.14) takes the linear form

$$g = \alpha + \beta r + \tau X/K,$$

the conditions for the existence of a meaningful equilibrium with excess capacity are

$$\alpha > 0, \quad [z/(1+z)](s-\beta) > a_1(\alpha+\tau).$$

Table 2.4. *Effects of parametric shifts in the Kalecki-Steindl model*

Parameter Variable	z	E	s	a_1	a_0
g	$-\dfrac{g_2 g/z^2}{s-g_1-g_2\frac{1+z}{z}}$ $(-)$	$\dfrac{s}{s-g_1-g_2\frac{1+z}{z}}$ $(+)$	$-\dfrac{g\left(g_1+g_2\frac{1+z}{z}\right)}{s\left(s-g_1-g_2\frac{1+z}{z}\right)}$ $(-)$	0	0
C	$-\dfrac{(1+s+z)}{(1+z)^2 a_0}$ $(-)$	0	$-\dfrac{z}{(1+z)a_0}$ $(-)$	0	$-\dfrac{[1+(1-s)z]}{(1+z)a_0^2}$ $(-)$
r	$-\dfrac{g_2 r/z^2}{s-g_1-g_2\frac{1+z}{z}}$ $(-)$	$\dfrac{1}{s-g_1-g_2\frac{1+z}{z}}$ $(+)$	$-\dfrac{r}{s-g_1-g_2\frac{1+z}{z}}$ $(-)$	0	0
W/P	$-\dfrac{1}{a_0(1+z)^2}$ $(-)$	0	0	0	$-\dfrac{1}{(1+z)a_0^2}$ $(-)$
X/K	$-\dfrac{1}{z^2}\left[\dfrac{g_2}{s-g_1-g_2\frac{1+z}{z}}\dfrac{1+z}{z}+1\right]$ $(-)$	$\dfrac{(1+z)/z}{s-g_1-g_2\frac{1+z}{z}}$ $(+)$	$-\dfrac{X/K}{s-g_1-g_2\frac{1+z}{z}}$ $(-)$	0	0
WL/rPK	$-\dfrac{1}{z^2}$	0	0	0	0

Note: The desired accumulation function is written as $g=g(r, X/K, E)$ with $\partial g/\partial r=g_1$, $\partial g/\partial (X/K)=g_2$, and $\partial g/\partial E=1$. See also the note in Table 2.1.

The second condition also implies that the responsiveness of saving exceeds the (total) responsiveness of investment, both to changes in the decision variable r, a standard condition that is required for macroeconomic stability (using dynamics such as those discussed in the next section). The first condition, would be violated under the plausible assumption (mentioned in section 2.2.3) that firms require a minimum rate of profit to invest anything at all. However, the condition is required only because we have assumed a linear investment function for deriving it; for a concave function in $\langle g, r \rangle$ space and $g(0, 0) < 0$, there could be two intersections of the investment curve with the saving curve as in the neo-Keynesian case of Figure 2.6, and the equilibrium with the higher r would be the stable one for a plausible adjustment process (see the next section). Even for a linear investment curve, the condition is not necessary in a more general version of the model.[15]

Among the comparative statics results – summarized in Table 2.4 – of greatest interest are effects of a change in z. A fall in z increases the real wage and causes a redistribution of income towards workers. Since the fall in z, for given r, increases $[(1 + z)/z]r$, the g function is shifted up in Figure 2.7, resulting in higher g^* and r^*. The wage-profit rate configuration moves closer to the potential wage-profit frontier, implying a higher degree of capacity utilization. The implications are that a better distribution of income accompanies a higher rate of growth, and a higher real wage is associated with a higher profit rate, given technology. That these results should be obtained seems somewhat paradoxical, and is a novel feature of this model compared to that of the three previous models considered; indeed they (especially the positive relation between wage and profit rate) run counter to established wisdom on the subject. Both results depend crucially on the assumption that the economy operates at less than full capacity in equilibrium. Starting from such an equilibrium, a lower z and a better income distribution increases consumption spending (with workers having a higher propensity to consume), raises demand, increases the rate of profit and capacity utilization, and raises the rate of accumulation. Changes in the parameters of the investment function and s have the same effects on growth as in the neo-Keynesian model, but they do not change the real wage. Instead of forced saving, they occur as a result of changes in the degree of capacity utilization.

2.3 The short-run and long-run dynamics

This section will look behind the long-run equilibrium framework adopted so far and examine the operation of the economy in the short run and in the long run when it is not in equilibrium. By so doing we will understand the mechanisms that take the economy to their long-run equilibrium positions, and also find that some closures other than the ones discussed in the previous section are possible.

The potential output, X_p, of the (representative) firm is

$$X_p = \min[N/a_0, K/a_1], \tag{2.18}$$

where N is the supply of labour. In the short run we assume that N and K are given, so that X_p is also given. The actual level of output in the short run is, in general, $X \leqslant X_p$, its level depending on the market structure of the goods market, and on the level of aggregate demand.

If the goods market of the economy is perfectly competitive, the price-taking firm will want to produce as much as it possibly can to maximize its profits, so that $X = X_p$. Given the endowments of N and K, the economy will be constrained by capital, labour, or both (if $N/K = a_0/a_1$). Once X is determined, the employment of labour can be determined from $L = a_0 X$; the unemployment rate is then determined. If the economy is labour constrained, we have full employment of labour but excess capacity. If it is capital constrained, $L < N$, and there is unemployment. If it is (accidentally) constrained by both, we have both full employment and capacity utilization.

With imperfect competition, since the firm has the power to fix its price, we must make some assumption about pricing behaviour. We assume, as in section 2.2.4, that the firm sets price according to (2.13) and changes the level of output in response to changes in demand if it can do so.

We maintain the assumption of price-taking behaviour for the neo-classical and neo-Marxian models, assume price-making behaviour for the Kalecki-Steindl model, and consider two versions of the neo-Keynesian one, one with perfect, and the other with imperfect competition. For each model we consider the short run (with N and K and perhaps other specified parameters given), and then examine convergence to long-run equilibria as examined above. The short-run behaviours we examine need not be the only ones consistent with the long-run equilibria considered above, but seem to be the simplest and most natural ones to consider.

2.3.1 The neoclassical model

For the neoclassical model we assume that, in addition to given K and N, the real wage, V, is given in the short run.[16] As noted above, perfectly competitive firms will produce capacity output X_p, so that labour employed will be $a_0 X_p$. The wage bill is then $V a_0 X_p$, and the profit level, $(1 - V a_0)X_p$, in real terms. With capitalists investing all their saving,

$$I = s(1 - V a_0)X_p. \tag{2.19}$$

It follows that consumption and investment demand always exhaust potential output: Say's law holds and there can be no effective demand problem.

The short-run values of all the variables are thus solved. X is equal to X_p, given by (2.18). Given V,

$$r = (1 - V a_0)X_p/K. \tag{2.20}$$

The initial endowments of N and K determine whether the labour or capital constraints will be binding. With $N/a_0 > K/a_1$ and the capital constraint binding, $X = K/a_1$, so that

$$r = (1 - Va_0)/a_1, \tag{2.21}$$

and there will be unemployment. With $N/a_0 < K/a_1$ and the labour constraint binding, $X = N/a_0$, so that

$$r = (1 - Va_0)N/a_0 K, \tag{2.22}$$

and there will be excess capacity; firms will wish to increase production by hiring more workers.

Beyond the short run, N, K, and V will change over time. We continue assuming that N grows at rate n. The growth rate of K, g, is given by $I/K = sr$. We assume that unemployment (excess demand for labour) results in a fall (rise) in V over time (due to competitive pressures or changes in relative bargaining strengths), which may be formalized by[17]

$$dV/dt = \Omega[(a_0 K)/(a_1 N) - 1]. \tag{2.23}$$

Figure 2.8 shows the dynamics. The $dV/dt = 0$ curve shows combinations of N/K and V which make $dV/dt = 0$. Equation (2.23) implies that it is vertical at $N/K = a_0/a_1$, and that to the left (right) of the curve V rises (falls), explaining the direction of the vertical arrows. The $d(N/K)/dt = 0$ curve shows combinations of N/K and V which make $d(N/K)/dt = 0$. To examine its shape note that

$$d(N/K)/dt = (n - g)N/K. \tag{2.24}$$

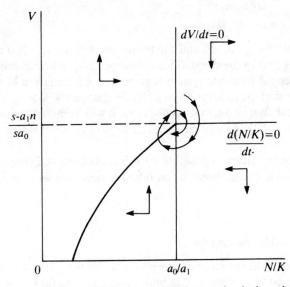

Figure 2.8 Dynamics in the neoclassical model

Equations (2.4), (2.21), and (2.22) imply

$$g = s(1 - Va_0)/a_1 \quad \text{for} \quad N/a_0 > K/a_1,$$
$$g = s(1 - Va_0)N/a_0 K \quad \text{for} \quad N/a_0 < K/a_1, \tag{2.25}$$

which implies, using (2.24),

$$d(N/K)/dt = [n - s(1 - Va_0)/a_1]N/K \quad \text{for} \quad N/K > a_0/a_1,$$
$$d(N/K)/dt = [n - s(1 - Va_0)N/a_0 K]N/K \quad \text{for} \quad N/K < a_0/a_1. \tag{2.26}$$

Thus, for $N/K \geqslant a_0/a_1$, $d(N/K)/dt = 0$ implies $V = (s - a_1 n)/sa_0$ so that the $d(N/K)/dt = 0$ curve is horizontal at this value of V, and for $N/K \leqslant a_0/a_1$, $d(N/K)/dt = 0$ implies $V = (1/a_0) - n/[s(N/K)]$ implying the upward rising segment. The direction of the horizontal arrows can be established from (2.26). The movement of the economy is seen to be cyclical, alternating between excess capacity and unemployment.[18] The (local) stability of this long-run equilibrium position can be shown as follows. The Jacobian of the dynamic system given by (2.23) and the first equation of (2.26), evaluated around long-run equilibrium, is

$$\begin{bmatrix} 0 & -\Omega(a_0/a_1) \\ s(a_0/a_1)^2 & 0 \end{bmatrix}$$

implying that its trace $= 0$ and determinant $= s\Omega a_0/a_1 > 0$. If the dynamic system were given by these equations, the equilibrium would be a centre. The Jacobian evaluated around the long-run equilibrium of the system given by (2.23) and the second equation of (2.26) is given by

$$\begin{bmatrix} 0 & -\Omega/(a_0/a_1) \\ s(a_0/a_1)^2 & -s(1 - Va_0)/a_1 \end{bmatrix}$$

implying a trace $= -s(1 - Va_0)/a_1 < 0$ and a determinant $= s\Omega a_0/a_1 > 0$. If the dynamic system were given by these equations the long-run equilibrium would be a stable focus. Since the dynamic system is given by a combination of the two systems, it follows that the system will be stable; in the zone $N/K > a_0/a_1$ it will be on a cyclical path, but in the zone $N/K < a_0/a_1$ it will be pulled towards the long-run equilibrium, falling into a circle nearer it the next time it enters the $N/K > a_0/a_1$ zone.

These are not the only dynamics possible for this model; we will consider alternative dynamics which can result in an (almost) neoclassical long-run equilibrium below.

2.3.2 The neo-Marxian model

For this model we assume the existence of a reserve army, so that $N/a_0 > K/a_1$. The real wage is assumed to be fixed at $\tilde{V}$ in both short and long runs; despite the existence of the reserve army, the real wage does not fall even in the long run.

In the short run, due to price-taking behaviour, $X = X_p = K/a_1$, so that employment is $a_0 K/a_1$. Each worker produces $1/a_0$, but is paid $\tilde{V}$, so that the surplus generated is $1/a_0 - \tilde{V}$. The rate of profit is given by

$$r = (1 - \tilde{V}a_0)/a_1. \tag{2.27}$$

Saving out of profits, srK in real terms, is entirely invested, implying, as in the neoclassical model, Say's law.

While the above analysis assumed that the real wage is always equal to $\tilde{V}$, we could allow it to deviate from $\tilde{V}$ in the short run. In this case our entire preceding analysis could be valid if we replaced $\tilde{V}$ by a given V. We could then append an adjustment equation

$$dV/dt = \Omega[\tilde{V} - V]. \tag{2.28}$$

It follows that medium-run equilibrium with $dV/dt = 0$, with K and N constant, implies $V = \tilde{V}$ and that the adjustment is stable. This type of wage adjustment, where the wage is pushed towards the 'subsistence' wage, should be distinguished from the neoclassical wage adjustment, where the real wage adjusts to the extent of slack in the labour market. We now return to the assumption of a fixed $\tilde{V}$.

Over time, N and K change. N increases at rate n, while the rate of growth of K is given by $g = sr$. Assuming that $n > g$ (so that there is never any shortage of labour) as K and N change over the long run, since $\tilde{V}$ is assumed given, the same short-run equilibrium is repeated over time.[19] If, on the other hand, $n < g$, the economy would eventually hit the full employment ceiling; the structure of the model would then need modification along the following lines. First, a_0 could be endogenized with capitalist firms introducing labour-saving technology in response to the labour shortage. Secondly, n could be endogenized, allowing capitalists to actively hunt for new sources of labour. Thirdly, $\tilde{V}$ could be endogenized as in the neoclassical model, yielding a neoclassical long-run equilibrium. Finally, excess capacity could be allowed to emerge in the long run by assuming that excess demand for labour pushes up the *money* wage, but the price rises at the same rate because firms keep their markup fixed (imperfect competition must enter the picture) so that the real wage does not change.[20]

2.3.3 The neo-Keynesian model with perfect competition

In the neo-Keynesian model with perfect competition we assume that N and K are fixed in the short run, and for simplicity the money wage, W, is fixed forever. We assume that the firms have a desired accumulation function. For simplicity we initially assume that the desired rate of accumulation, g, is given, so that investment I, equal to gK, is given in the short run.

Equilibrium in the short run requires $X = C + I$, so that

$$X = (W/P)a_0 X + (1-s)[X - (W/P)a_0 X] + I. \tag{2.29}$$

For any given X, we assume that the price clears the market in a Walrasian manner. The market-clearing price is then

$$P = (sWa_0 X)/(sX - I). \tag{2.30}$$

The curve showing the relationship between X and P given by this equation, which has a negative slope $-(sWa_0 I)/(sX - I)^2$, is the aggregate demand curve (*AD*) in Figure 2.9.

With perfect competition, $X = X_p$. For given K and N in the short run, there are two possible cases, $N/a_0 > K/a_1$ and $N/a_0 < K/a_1$. In the first case, we have $X = K/a_1$; from (2.30), substituting this X and $I = gK$ we have $W/P = (s - a_1 g)/sa_0$. In the second case, $X = X_p = N/a_0$ (there is an excess capacity of capital) and $W/P = (sN/Ka_0 - g)/(sN/K)$. In both cases, since $g = sr$ we have $r = g/s$.

In the long run, starting from given initial levels of K and N the economy moves over time with K and N increasing at rates g and n, respectively. In the neo-Keynesian model of section 2.2.3, $g = I/K$ depends only on r, but with the possibility of excess capacity arising due to labour shortage, we assume here that it depends also on the rate of capacity utilization (as in the Kalecki-Steindl model). Adopting a linear form,

$$I/K = \alpha + \beta r + \tau(X/K), \tag{2.31}$$

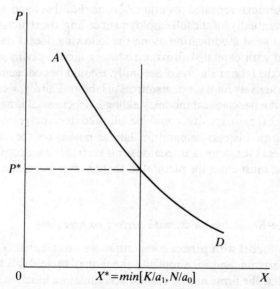

Figure 2.9   The neo-Keynesian model with perfect competition: the short run

where α, β and τ are all positive. In the capital constrained case, with $X/K = 1/a_1$, this implies

$$I/K = \alpha + \beta r + \tau/a_1 \tag{2.32}$$

while in the labour constrained case, with $X/K = N/a_0 K$, we have

$$I/K = \alpha + \beta r + \tau(N/a_0 K). \tag{2.33}$$

In short-run equilibrium the goods market must clear, so that $S/K = I/K$. In the capital constrained case (2.4) and (2.32) imply

$$r = [\alpha + \tau/a_1]/(s - \beta),$$

so that (2.4) yields

$$g = s[\alpha + \tau/a_1]/(s - \beta). \tag{2.34}$$

Similarly, for the labour constrained case, (2.4) and (2.33) imply

$$r = [\alpha + \tau(N/a_0 K)]/(s - \beta),$$

so that

$$g = s[\alpha + \tau(N/a_0 K)]/(s - \beta). \tag{2.35}$$

Figure 2.10 shows the dynamics for this model. The horizontal line g^K shows the value of g for each N/K in the capital-constrained case, given by (2.34), while the upward-sloping line g^N shows the value of g for each N/K in the labour-constrained case, given by (2.35); by the g curve we will denote the segment of the g^N curve for the excess capacity region with $N/K < a_0/a_1$, and the segment g^K for $N/K > a_0/a_1$, the two being equal at $N/K = a_0/a_1$. The figure also shows the rate of growth of N, n. Two cases must be distinguished, depending on whether a_0/a_1 is less than or greater than the N/K at which the n and g curves intersect.[21] Figure 2.10(a) shows the first case, which results in a long-run equilibrium identical to the one examined for the neo-Keynesian model of section 2.2.3, with the economy growing with full capacity following a desired accumulation function where the rate of accumulation depends only on r. Whether it starts from an initial position of excess capacity (with $N/K < a_0/a_1$) or full capacity (with $N/K > a_0/a_1$), the economy will move along the arrows (hitting full capacity if it started off with excess capacity) and eventually move along the g^K line with increasing N/K and unemployment. Figure 2.10(b) shows the second case, in which whether we start with excess capacity (with $g < n$, below E or with $g > n$, between E and a_0/a_1), or with full capacity (above a_0/a_1), we end up at E, in a position of long-run full employment equilibrium with excess capacity.

This equilibrium, different from all the others we have considered, is given by equations (2.1), (2.2), (2.4), (2.6), and (2.31), which is a special case of (2.14). This closure, because it is similar to that of the full employment models of Kaldor (1955) which combine investment functions with the full employment

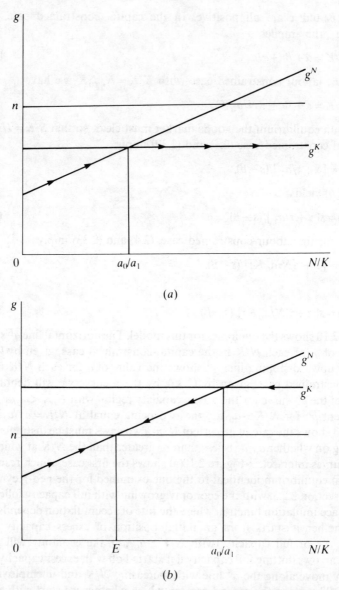

Figure 2.10 Dynamics in the neo-Keynesian model with perfect competition

assumption, can be called the Kaldorian one. The model has the property that,
if there is an increase in some of the investment parameters, greater excess
capacity will emerge in long-run equilibrium. If the presence of excess capacity
in long-run equilibrium creates pressures for these parameters to fall,

eventually the g and n curves will intersect at a_0/a_1, yielding the neoclassical long-run equilibrium with full capacity utilization.

One final remark is in order. In the short run we have made the goods market clear by variations in P, given W. Suppose, however, that there is a subsistence real wage $\tilde{V}$ which must be met, and excess demand in the goods market cannot be cleared in the short run by increases in P since the subsistence real wage has been reached. In this case we need some other adjustment mechanism. Excess demand can be rationed by reductions in consumption out of profits, or of investment, or both (with no reductions in subsistence wage consumption possible). If investment has to adjust, we are back in the neo-Marxian model (since the investment function no longer plays a role in the model), but if the adjustment falls on s we have a different type of closure. In this case equilibrium is given by equations (2.1), (2.3), (2.4), (2.5), (2.8), and (2.10), where s now becomes an additional variable, as in models suggested by Johansen (1960), in which s changes due to fiscal policy changes.[22]

2.3.4 Models with price setting behaviour

In this case we assume that firms set the price according to the Kaleckian formula (2.13) if they can adjust output when demand changes, but when upward quantity variations are not possible (at X_p) they allow the price level to clear the market. One way to formalize this is to assume that firms have a supply price, P^s, given by

$$P^s = (1+z)Wa_0. \tag{2.36}$$

The degree of monopoly power affects the level of z and hence P^s. The demand price, for any *given* X, defined as the price which clears the market for that output level, is

$$P^d = sWa_0 X/(sX - I). \tag{2.37}$$

For a given X, the price is determined at P^d. Firms then adjust output according to

$$dX/dt = \Omega[P^d - P^s] \tag{2.38}$$

for $X < X_p$, which shows that they expand output if the demand price exceeds the supply price and conversely if it is less.[23] This implies that for $X < X_p$, short-run equilibrium, requiring $dX/dt = 0$ requires

$$P^d = P^s. \tag{2.39}$$

Equations (2.36), (2.37), and (2.39) imply

$$X = (1+z)I/zs. \tag{2.40}$$

Since this case is valid for $X < X_p$, we require I, z, and s to satisfy

$(1 + z)I/zs < X_p$. If this is not satisfied, so that $X = X_p$ (set either by the labour or capital constraints), X cannot rise with $P^d > P^s$, so that $P = P^d$ at the given X; the equilibrium is the same as the one of section 2.3.3. Figure 2.11 depicts all this.

For the case where the equilibrium X is less than X_p, we have, substituting $I = gK$ into (2.40),

$$X/K = (1 + z)g/zs. \tag{2.41}$$

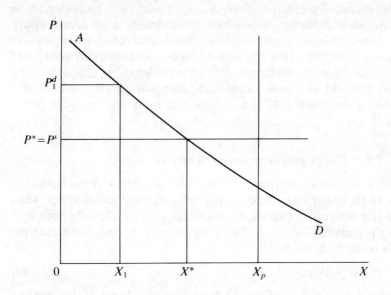

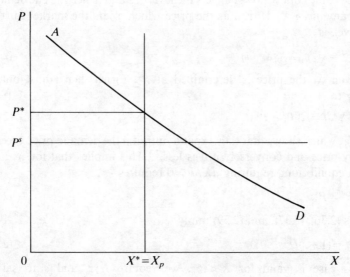

Figure 2.11   Short-run equilibrium with price-setting firms

Given $g = sr$ (which must be satisfied since in short-run equilibrium the goods market clear), we also have

$$r = z/(1+z)X/K. \tag{2.15}$$

If we assume g to be (provisionally) given, the short-run (for given K) equilibrium values of X and r are found from these equations. In the long run, K changes at rate g. Inspection of the last two equations shows that the system is also in long-run equilibrium, since changes in K will change X equi-proportionately, leaving X/K, r, and the other variables of the model unchanged. The case of the more general investment function (2.31) can be considered in a similar manner.[24]

For the full capacity cases in which $X = X_p = K/a_1$ or $X = X_p = N/a_0$, the equilibria are the same as in the neo-Keynesian and Kaldorian closures, respectively, considered in section 2.3.3. The only difference is in the mechanism which drives the economy to full capacity utilization: while it was price-taking behaviour in the earlier analysis, here it is high level of demand in an imperfectly competitive environment.

Depending on the parameters of the mode, the economy could arrive at one of the three possible equilibria. With a sufficiently high markup, and with a long-run equilibrium $g < n$, the economy will be in a Kalecki-Steindl equilibrium; if equilibrium $g > n$, it could change regimes to a Kaldorian equilibrium. With a low markup, the neo-Keynesian or Kaldorian closures could apply. Starting from either the neo-Keynesian or Kaldorian long-run equilibrium, an exogenous increase in z could shift the economy to a Kalecki-Steindl equilibrium.

2.4 Conclusion

This chapter has introduced the neoclassical, neo-Marxian, neo-Keynesian, and Kalecki-Steindl models, which will be used in the rest of the book for studying a variety of issues. They will be referred to below as the basic models.

They have been presented as alternative closures of a basic framework for analysing the determination of growth and income distribution in capitalist economies. Some additional closures have been mentioned, but will not be given pride of place alongside these four.

The models have been developed using a variety of restrictive assumptions, and little effort has been made to justify the names given to them. The next chapter will relax some of these assumptions, and by so doing and by placing them in the theoretical contexts to which they belong, seek to justify the names we have given them.

CHAPTER 3

Alternative models and alternative approaches

3.0 Introduction

The previous chapter examined (at least) four different models representing four important traditions analysing growth and distributional issues, but no attempt was made to justify their representativeness of the alternative traditions. There is need for such justification, however, since those belonging to a particular tradition may deny that the model which is supposed to represent their approach does so adequately. While this may partly be the irrational but understandable reaction to being told that what one considers to be the only theory is one of several special cases of a more general approach, the problem may be a real one, for several reasons.

First, the different approaches could address entirely different kinds of questions and answer them at different levels of analysis. The Marxist tradition may be interested mainly in how different social institutions such as capitalism may be transformed, and how this in turn changes legal, social, political, and other aspects of society, and the neoclassical one may primarily be interested in how a free enterprise economy allocates scarce resources for the 'common good'. Answering such different questions requires different levels of analysis: the former requires an extremely broad canvas, but the analysis must necessarily be somewhat general, while the latter is narrower in scope and the analysis can afford to be far more precise. Our comparative approach forces each tradition to ask the same questions and use the common tool of simultaneous equations. While the questions addressed by the different approaches may all be important and one need not be rigid about one's analytical tools, our defence is that our method – as long as it respects the existence of others – cannot hurt, and may well help in the analysis of the issues which concern us here.

Secondly, different approaches sometimes have different organizing principles which, as discussed in chapter 1, provide unifying meta-assumptions which are not meant to describe the real world, but allow all explanation of relevant phenomena to be reduced to them. Our comparative framework, forsaking all such organizing principles, uses definitional equations and plausible behavioural stories. While different organizing principles in practice may determine what questions are asked and what techniques of analysis are chosen within a particular tradition, we have argued that organizing principles need not make a real difference; moreover, their use without understanding their role in theory can be dangerous. To illustrate some of these issues we will employ some of the organizing principles in the models for the traditions in which they are commonly used, and also explore how some models not usually using a particular organizing principle may have to be modified by its use.

Thirdly, and perhaps most important, our general framework may be too specialized, already ruling out some features of the economy which are important for the different traditions. Our framework assumes fixed coefficients of production, while neoclassical economics usually assumes smooth substitution in production. We have assumed away technological change, which plays a central role in the Marxian approach. Monetary issues, despite their importance in Marxian and Keynesian theory, have been abstracted from. Perhaps no manageable framework can be general enough to do justice to any of the traditions. This chapter, however, relaxes some of the assumptions made so far to make the models become more recognizably representing their tradition.

We now examine each of the approaches in turn, justifying the identification of our models with each of the traditions, and modifying some of the assumptions of the previous chapter. The main modifications – which are made here for illustrative purposes, rather than in a systematic manner – involve the introduction substitution in production, optimizing behaviour, and some implications of class struggle; extensions due to technological change, monetary issues, multi-sector complications, and open-economy considerations are postponed to later chapters.

3.1 The neoclassical approach

3.1.1 The neoclassical approach and closure

The neoclassical approach represents the mainstream of the economics profession today. Since the practitioners of neoclassical economics typically believe that theirs is the only legitimate kind of economics, and do not think of themselves as belonging to a particular school, it is somewhat difficult to define what is meant by neoclassical economics. An examination of their work, however, seems to suggest that they (usually) believe in the following:

1 *Principle of optimizing agents.* Economic behaviour should be examined explicitly in terms of the behaviour of agents who optimize subject to the constraints they face.[1]
2 *Strong perfect markets principle.* The economy is characterized by relatively few rigidities and imperfections, so that the appropriate model of an economy to use is one in which perfect competition prevails in all markets which are cleared by instantaneous price movements.

3 *Weak perfect markets principle.* Even if the economy is not perfectly competitive, all analysis of economic systems should treat perfectly competitive equilibrium as a benchmark, since under certain conditions it represents a Pareto optimal outcome which, distributional issues aside, is a desirable outcome; distortions in an economy, if they exist, can be thought of as representing deviations from the benchmark and policy issues can be discussed as ways of returning to that state.
4 *Principle of exogeneity of non-economic factors.* Non-economic factors such as political, sociological, psychological, and technological ones, can be taken as given and outside the scope of economic analysis.

Not all neoclassical economists adhere to each of them. The strong perfect markets principle is probably believed by very few neoclassical economists,[2] and the exogeneity of non-economic factors principle (henceforth the exogeneity principle) has been modified by the endogenization of factors not traditionally considered to be economic (technological ones are the most popular).[3] The optimizing agents and weak perfect markets principles are probably the ones most commonly followed, although there are notable examples of eminent neoclassical economists not adhering to either.[4] But most neoclassical economists would probably admit to using at least three of these four principles (excepting the strong perfect markets principle), and some would agree to all (although once (2) is accepted, (3) becomes redundant).

Our neoclassical model, however, seems to be using only two of the principles, those of strong market clearing and exogeneity, and, by threatening to consider technological change in a later chapter, seems faithful only to the former. Yet this is the characteristic which most neoclassical economists would be willing to drop.[5] How can we then call ours neoclassical?

The optimizing agents and weak perfect markets principles, especially the former, may be regarded as providing the organizing principles of neoclassical economics. We have argued in chapter 1 that these are not principles which truly distinguish economists of different perspectives, but what economists have found useful to organize their analysis of particular questions around. There is no real reason why the analysis of a particular question using alternative organizing frameworks must yield different answers, since they do not involve assumptions about the real world. If this is true, non-neoclassical economists need have no real quarrel with the use of such organizing principles.[6] They could, however, object to particular uses of the optimizing framework (ignoring crucially important constraints or misspecification of the general environment in which optimization is carried out, a criticism, of the strong perfect markets principle, invariably),[7] or the claim that all legitimate analysis *must* use optimizing behaviour (or in the analysis of economic systems, the Pareto optimality criterion) as the organizing principle of analysis. In the analysis of economic growth questions, the Pareto optimality notion has been less commonly adhered to by neoclassicals than in the analysis of static issues. This is partly due to the non-existence of infinitely lived optimizing individuals (overlapping generations have not been able to come to the rescue),[8] and partly due to the endogeneity of preferences in terms of which Pareto optimality can be defined (this is really a criticism of the exogeneity principle). The optimizing principle, however, remains a potentially useful tool,[9] and will be explored partially for the neoclassical (and other) models.

The exogeneity principle involves the use of a particular dichotomy, as defined in chapter 1. Neoclassical economists can and have been criticized by others for using it. Our neoclassical model employs this dichotomy (since perhaps s depends on psychological or sociological factors,[10] the parameters of the production function are given technologically, the rate of population growth is given perhaps sociologically, and the government is ignored). But we prefer to ignore this criterion in identifying neoclassical models, because: as already noted above, many neoclassical economists endogenize non-economic features; many non-neoclassical economists have used similar dichotomies; and models employing these dichotomies can, if necessary, be extended to endogenize the non-economic variables,[11] so that the dichotomy is not intrinsic to the neoclassical.

We are thus left with only the strong perfect markets principle, which in our simple model implies perfect competition in the goods market and market clearing (no unemployment) in the labour market. The main reason for using this characteristic as the distinguishing feature of our neoclassical model is that in the historical development of the alternative approaches, the full employment assumption has been the hallmark of the neoclassical approach.[12] The Solow-Swan models which pioneered the approach in the growth theory literature used the assumption, and it is precisely that feature which distinguishes it from the other approaches. Also, while in short-run macroeconomic theory neoclassical economists (apart from the monetarists and rational expectations theorists) have often forsaken the assumption of full employment, they assume full employment in the long run, which is precisely the sense in which full employment is assumed in our model.[13]

3.1.2 Modifications of the neoclassical closure

While we have explained why we have identified our full employment closure with the neoclassical approach, we must admit that there are several assumptions made that are contrary to usual neoclassical assumptions. First, there is the assumption of two classes while neoclassical models such as Solow's (1956) consider only one class. Our model is actually closer to the models of Kaldor (1955–6) and Pasinetti (1962), which are usually identified with the neo-Keynesian approach, but which assume full employment growth in the long run.[14] Secondly, we assume fixed coefficients technology, while neoclassical economists usually allow smooth substitution in production.

If, ignoring class distinctions, we assume that all income goes to households who save a constant fraction σ (to distinguish this from s, the saving rate out of capitalist income) of their income, we have

$$S = \sigma X, \tag{3.1}$$

where S is real saving. Assuming fixed coefficients as before, we get $S/K = \sigma/a_1$, and, since all saving must be invested, $g = \sigma/a_1$. This is the rate of growth of capital stock (and total product) which is consistent with the saving behaviour of the economy and full capacity. If full employment growth is desired we require

$$\sigma/a_1 = n. \tag{3.2}$$

However, the three parameters in the equation are independently fixed, σ perhaps by sociological and psychological factors, n by demographic factors, and a_1 by technological factors, and there is no reason they should satisfy (3.2). This problem, known as Harrod's long-run problem, is supposed to demonstrate that capitalist economies generally exhibit unemployment (if $n > \sigma/a_1$) or excess capacity (if $n < \sigma/a_1$), and growth with full employment and full capacity can only be accidental.

There are several ways of showing that the problem arises only because of the rigidity of Harrod's (1939) assumptions. One 'solution', provided by the Kaldor-Pasinetti model, is to endogenize σ by assuming that there are two classes with two different saving propensities, and by letting the distribution of income between the two groups change endogenously. Kaldor (1955–6) considered two types of income, wage and profit income, and assumed saving from each type of income to be a constant fraction of that type of income, with the fraction relating to profit income being higher than that of wage income. Pasinetti (1962) altered the categories to workers and capitalists, and allowed both workers and capitalists to save, with capitalists having a higher propensity to save.[15] Our neoclassical model simplifies the Pasinetti approach, by assuming that workers do not save.

Another approach is to endogenize n. Malthusian assumptions could be made which raise the rate of population growth when $\sigma/a_1 > n$ and reduce it when $\sigma/a_1 < n$, so that movements in n would ensure full employment in long-run growth.

But these are not the typical neoclassical solutions, which is rather to allow individual optimization to play some role, or allow for some substitution in production.

Individual optimization could be given some role by assuming that individuals maximize lifetime utility in making their saving decisions. A simple and elegant model assumes overlapping generations, with each individual living for two periods – one for working, earning and saving, and the other for retiring and dissaving.[16] The substantive difference made by this modification is that the aggregate saving-income ratio σ now depends on the rate of interest or profit. If, and only if, σ rises with r, the economy converges to full employment growth.

Solow's (1956) neoclassical model, however, replaced fixed coefficients with the assumption of smooth substitution in production. Following him, consider the production function

$$X = F(K,L) \tag{3.3}$$

satisfying the usual neoclassical properties of differentiability of desired order, positive marginal products, diminishing returns to factors, and constant returns to scale. The representative, perfectly competitive, firm is assumed to maximize profits, and therefore employ labour and capital to equate their real costs to their marginal products.

Assume that in the short run K and L are fixed. Then real wage and rental rate variations will clear the markets for both factors in the short run. We assume that capital here is either hired, or that it is financed by borrowing and with perfect competition the cost of borrowing is equal to the rate of profit. The short-run equilibrium is shown in Figure 3.1: with K_0 and L_0 as endowments, equilibrium occurs at E, production (which, by Say's law is sold) is X_0 as shown by the isoquant, and the wage-rental ratio is given by the slope of the isocost line AB tangent to the isoquant at E. Notice that firms hire all capital and labour.

Beyond the short run, Solow's model assumes that N (which is always equal to L) grows at the rate n, as assumed before. Given the fixed saving rate

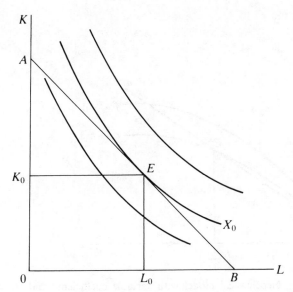

Figure 3.1 Neoclassical model with variable coefficients: the short run

σ capital grows at the rate

$$g = \sigma X/K. \tag{3.4}$$

Depending on the values of σ, n, and the initial endowments of K and L (which determine X/K), g will be greater or less than n. If $g > n$, for instance, capital grows faster than labour, so that (see Figure 3.1) the wage-rental ratio rises, inducing the firm to substitute capital for labour, increasing K/X, and hence reducing g until it becomes equal to n. The opposite would occur with the wage-rental ratio falling if $g < n$. In either case, the economy would eventually tend to a position with $g = n$, satisfying (3.2).

All this can easily be shown using Solow's original diagram in Figure 3.2. Given our assumptions, it is well known that we can rewrite (3.3) as

$$1/a_0 = f(a_1/a_0), \tag{3.5}$$

where $1/a_0$ is output per unit of labour, and a_1/a_0 is the capital-labour ratio. Since we now have variable proportions, a_0 and a_1 are no longer fixed, but satisfy $1 = F(a_1, a_0)$. It is easy to show that $f' > 0$ and $f'' < 0$, with f' being the marginal product of capital, and $f - (a_1/a_0)f'$ the marginal product of labour. Our assumptions imply that

$$r = f', \tag{3.6}$$

$$V = f - (a_1/a_0)f'. \tag{3.7}$$

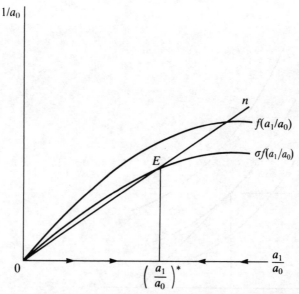

Figure 3.2   Neoclassical model with variable coefficients and constant saving-income ratio: the long run

The figure shows functions f, sf and na_1/a_0. We have

$$d/dt\,(a_1/a_0) = (g-n)\,(a_1/a_0) = \sigma f - n(a_1/a_0), \tag{3.8}$$

so that in the figure a_1/a_0 is rising for values less than that at E and falling for those above. Long-run equilibrium, with constant a_1/a_0, occurs at E when $\sigma f = (a_1/a_0)n$ which, using (3.5) satisfies (3.2), so that the economy grows with full employment and full capacity.[17]

This suggests that it is simple to modify our two-class neoclassical model to allow for variable coefficients of production. In the short run, with given K and L, output X, and hence the a_0 and a_1, are determined as in Figure 3.1, and r and V are determined from (3.6) and (3.7). The rate of growth of capital is now given by equation (2.4) and N is assumed to grow at rate n as before. Beyond the short run, the capital-labour ratio, a_1/a_0, changes according to

$$[1/(a_1/a_0)]d(a_1/a_0)/dt = sf' - n. \tag{3.9}$$

Convergence to long-run equilibrium with $g=n$ is shown in Figure 3.3.[18] Long-run equilibrium is now determined by

$$1 = Ca_0 + ga_1,$$
$$1 = Va_0 + ra_1$$

which are exactly the same equations as those of the neoclassical model of the last chapter. However, while a_0 and a_1 are fixed in the earlier model, here they

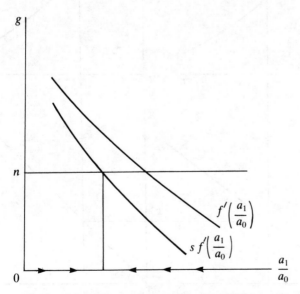

Figure 3.3 Neoclassical model with variable coefficeints and differential saving propensities: the long run

respond to changes in factor prices. We saw that the slope of the wage-profit frontier, with fixed coefficients, is given by a_1/a_0. Now, as we move down the frontier, the wage-profit ratio falls, making the firms reduce a_1/a_0; thus the curve is convex to the origin, the extent of curvature depending on the elasticity of substitution. The consumption-growth frontier must take the same shape as the wage-profit frontier. The rest of Figure 3.4 is the same as Figure 2.3. An examination of the results of parametric shifts is not considered here, but are similar to those for the model with fixed coefficients.

3.2 The neo-Marxian approach

3.2.1 *The neo-Marxian approach and closure*

The neo-Marxian approach is far more difficult to delineate than the neoclassical one, because of the greater diversity among those, who because they derive primary inspiration from Marx, we may call neo-Marxian. A set of

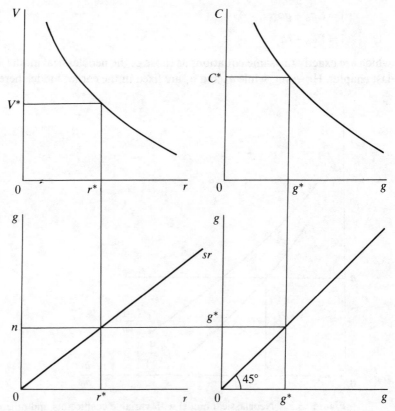

Figure 3.4   Neoclassical model with variable coefficients: long-run equilibrium

criteria to distinguish the neo-Marxist's approach may include the following: (1) analysis of economic process in terms of the interaction between the forces of production and the social relations of production, giving primary attention to the material base of the economy; (2) belief in the usefulness of some version of the labour theory of value; (3) belief in eventual transition from capitalism to socialism; and (4) belief in the principle that the distribution of income is best analysed in terms of social, political, historical, and moral factors which are exogenous to the narrowly defined economic forces of demand and supply. Although not all Marxists accept all of these canons, each one is sufficiently popular. We examine the role given to each in our model.

The first one, shared in some form by all neo-Marxists, can be called their organizing principle. It may be expressed in its popular form as 'all history is the history of class struggle'. Analysis using such an organizing principle may be extremely useful for the analysis of structural transformations in economies, but, in line with our usual practice of playing down the role of organizing principles, we do not emphasize this aspect, primarily because we are concerned here mainly with capitalist development and not structural transformations of economies from one mode to another. However, some vestiges of the idea remain in our analysis. The importance of the class distinction obviously reflects the influence of Marx's analysis of the social relations of production in capitalism: capitalists own the means of production, and workers who do not, are forced to sell their labour power for a wage. We shall also introduce the concept of the social determination of what we have called the technological parameters, which is in line with this organizing principle. Moreover, our analysis can lead the way to the analysis of the broader issues of structural transformation by examining parametric changes in the model: one example is technological change, which will be considered in chapter 5.

We have so far not mentioned the labour theory of value, since we have not been concerned with the 'source' of profit, but with its determination, which some Marxists call 'surface' phenomenon. Many Marxists, in any case, have come out openly against the labour theory of value in any form.[19] It is, however, a simple matter to examine some of the Marxian value categories in terms of our simple model, and will be done in this chapter.

The question of the downfall of capitalism cannot be properly dealt with here, partly because our analysis cannot address questions relating to endogenous political changes, and partly because we are using the framework of long-run equilibrium. Some issues, including those relating to the tendency of the rate of profit to fall, can be analysed only after we consider technological change in chapter 5. Some questions could be raised using our framework, however. Our model determines g in long-run equilibrium with $g < n$, implying that the reserve army of the unemployed increases over time. One can ask whether this is a politically stable development, or one which will engender revolutionary forces and lead to the fall of capitalism.

The element we have emphasized in our model is the idea of the exogeneity of income distribution. The specific form in which we have introduced this into our model is with the assumption of a subsistence wage, broadly defined. The subsistence wage idea has roots in earlier classical writers such as Ricardo and Malthus, where the wage tended to that level due to population dynamics. For Marx it was the existence of the reserve army of unemployed – due perhaps to the destruction of precapitalist modes of production, or the proletarianization of the peasantry – which pushed the wage down to the level determined by social, moral, and historical forces. An alternative assumption formalizing this Marxian exogeneity principle makes the wage *share* rather than the real wage *rate* exogenous. If class struggle is thought to be waged over the share in output, it is more natural to make this assumption, and we will explore its implications below.

3.2.2 Modifications of the neo-Marxian closure

Here we consider five issues related to our neo-Marxian model, and for some of them, examine simple modifications of the model.

First we introduce Marxian labour value concepts and the rate of exploitation. In the neo-Marxian model above (section 2.3.2) each worker produces $1/a_0$, but is paid $\tilde{V}$, so that $1/a_0 - \tilde{V}$ is the surplus generated. This, which may be called surplus value in terms of the good, may be translated into $1 - \tilde{V}a_0$, surplus value in terms of labour, or surplus labour. Since we abstract from intermediate inputs and depreciation (which made up Marx's 'constant capital' because it did not augment value), surplus labour is thus total labour less 'variable capital' or what is paid to workers. The rate of exploitation, ϵ, is thus

$$\epsilon = (1 - \tilde{V}a_0)/\tilde{V}a_0. \tag{3.10}$$

With the rate of profit given by (2.27) we get

$$r = \epsilon \tilde{V}a_0/a_1 \tag{3.11}$$

which shows that the necessary and sufficient condition for a positive rate of profit is a positive rate of exploitation. In this sense it may be claimed (see Morishima, 1973) that in the Marxian system, exploitation (which implies $\epsilon > 0$) is the source of profit. From (3.10) it also follows that, given the technological parameters, the given real wage implies a given rate of exploitation and vice versa. Thus, if we extend the neo-Marxian model to include equation (3.10), V can be interpreted as a variable to be determined once ϵ and a_0 are known, or ϵ can be the variable which is determined, given $\tilde{V}$ and a_0. Fixing ϵ or V are equivalent for given technology; when technology changes the two have very different implications, as we shall see in chapter 5.

Many neo-Marxists reject the use of labour values and the usefulness of using the rate of exploitation as a parameter, because one can understand the determination of growth and distribution (and, in multisector models, prices as well) without using them at all. Some Marxists, however, like to understand the origin of profits in terms of exploitation, measuring the rate of exploitation as a ratio of magnitudes expressed in terms of labour values. Some – Joan Robinson among them – have dismissed this as metaphysical, but as Sen (1978) argues, the labour theory of value and the concept of the rate of exploitation serve descriptive purposes. One can also argue that ϵ is a more fundamental parameter to focus on in considering the state of the class struggle, since both $\tilde{V}$ and a_0 (see the next paragraph) reflect the balance of class relations. An increase in the length of the working day could reduce a_0 and increase ϵ without changing $\tilde{V}$; this would reflect a worsening in the position of workers, but would not be captured in changes in $\tilde{V}$. However, when technological improvements which do not 'hurt' workers but raise labour productivity occur, $\tilde{V}$ may be the better indicator of class struggle.[20]

Secondly, we briefly touch on the social determination a_0 and a_1, which we have called the technological parameters. A truly Marxist analysis would find these parameters to be determined by socio-political forces as well as technological ones; they would reflect the state of class struggle. It may be argued that a_0, for instance, would depend on the length of the working day (if a_0 is interpreted as the number of workers needed per unit of output) as well as the intensity of work of the labourers. The former would depend on custom and political forces. The latter would reflect the power relations within the capitalist firm. Our model could be modified to take these effects into account, following Kalecki's (1971) argument that it is in the interest of capitalists to maintain unemployment for disciplining workers, and work by Bowles (1985) and others which shows that firms may want to raise wages to extract more effort. Bowles and Boyer (1988) have examined these issues using a macro-economic model.[21]

Thirdly, to evaluate the claim that the generality of the neo-Marxian model is marred by the fixed coefficients assumption,[22] we consider what may be called a neoclassical version of the neo-Marxian model with substitution in production and profit maximization. Using the smooth neoclassical production function given by (3.5) we note that our assumption of constant returns to scale implies (by the adding up theorem and the wage equals marginal product of labour condition) that the marginal product of capital would be equal to the rate of profit. Maintaining the other neo-Marxian assumptions, in particular, the exogeneity of $\tilde{V}$, we find that (3.7) determines the capital-labour ratio a_1/a_0, and (3.5) determines a_0, so that a_1 is also determined. Thus, even without fixed a_0 and a_1, the fixity of V in effect fixes them, so that the determination of equilibrium for the model can then be analysed in exactly the same manner as was done for the fixed coefficients case. The only difference would be in examining the effect of changes in $\tilde{V}$; in this model a rise in $\tilde{V}$ would induce firms to substitute capital for labour, reducing a_0 and increasing a_1.

Fourthly, we examine the implications of the fixed real wage assumption. Our neo-Marxian model – which fixes the real wage or the rate of exploitation – can be criticized as deviating too far from Marx's (1867) view that 'the rate of wages [is] the dependent, not the independent variable'.[23] We should stress that our assumption of an exogenous wage (or rate of exploitation) does not imply that in a broader model it must be considered exogenous; our crucial assumption is it does not vary to 'clear' the labour market to result in full employment growth. The exogeneity assumption provides a model which can be used for analysing the effects of systematic changes in the real wage over time. To illustrate this, we examine two ways of endogenizing the real wage in our neo-Marxian model, providing different results.

In the first one,[24] we assume that the real wage rises when the demand for labour grows faster than its supply and vice versa (due to changes in the state of the class struggle resulting from changes in the state of the labour market), so that in linear form,

$$dV/dt = \Omega[g - n],\tag{3.12}$$

where $\Omega > 0$ is an adjustment constant. This model is depicted in Figure 3.5, where the n line shows the given rate of population growth and the gg line shows the inverse, linear relation between g and V (see Table 2.2). Starting with $V > V^e$ ($V < V^e$) where $n > g$ ($n < g$), (3.12) implies that V falls (rises), proving convergence to V^e, the equilibrium real wage with $g = n$. Note that while the equilibrium condition for this model is the same as that of the neoclassical model, given by equation (2.7), it will not hold here except in the 'very' long-run equilibrium, and also it does not necessarily imply full employment, since it

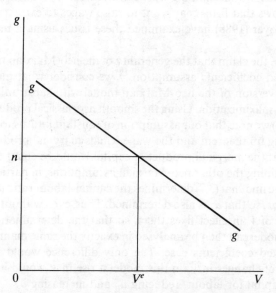

Figure 3.5   A neo-Marxian model with endogenous real wage

could imply only a given rate of unemployment. Thus, while this model shares the comparative dynamic properties of the neoclassical model at steady state, the two models are not identical.

A second version endogenizes the wage following Goodwin (1967). We assume that the change in the real wage depends on the rate of employment, $l = L/N$. Using a linear approximation, we have

$$\tilde{V} = -A + Bl, \tag{3.13}$$

where hats over variables denote rates of growth. We continue assuming that the rate of growth of labour supply is fixed at n and the input-output coefficients are given.[25] For simplicity, we also assume that $s = 1$ (capitalists do not save).

We follow Goodwin (1967) by examining the behaviour of the employment rate, l, and the wage share, $\delta (= Va_0)$. For δ, we have

$$\hat{\delta} = \hat{V}$$

which, using (3.13), implies

$$\hat{\delta} = -A + Bl. \tag{3.14}$$

For l, we have

$$\hat{l} = \hat{X} - n.$$

Since a_1 is constant, the rate of growth of X is equal to that of K. The rate of profit is given by $(1 - \delta)/a_1$, and is equal to the rate of growth of capital, since all profits are saved. Thus we have

$$\hat{l} = (1 - \delta)/a_1 - n. \tag{315}$$

Equations (3.14) and (3.15) yield

$$\dot{\delta} = [-A + Bl]\delta, \tag{3.16}$$

$$\dot{l} = \{[(1/a_1) - n] - (1 - a_1)\delta\}l \tag{3.17}$$

which are in the Lotka-Volterra form,[26] implying that δ and l are caught in a limit cycle. Since the growth rate of the economy is given by $g = (1 - \delta)/a_1$, the economy exhibits cyclical growth. It thus appears that, with a slight modification in the way in which the real wage is endogenized, we obtain a model with cyclical growth.[27] Notice that at any point on the cycle all of the equations of our neo-Marxian model are satisfied: thus the fact that most of this book is concerned with steady-state behaviour should not lead us to conclude that the approach is inconsistent with cyclical, or even unstable, behaviour of the economy.[28]

Finally, we comment on the role of demand in Marx's analysis. While our neo-Marxian model ignores demand, Marx departed from the Ricardian idea that there can never be an over supply of goods – a notion later to be called Say's law. He argued that, due to the anarchical nature of production (with uncoordinated decisions made by many producers) and the existence of money (which separated the act of sale from the act of purchase), crises of overproduction were possible.[29] Marx (1894) wrote:

> The creation of ... surplus value is the object of the direct process of production ... But this production of surplus value is but the first act of the capitalist process of production ... Now comes the second act of the process. The entire mass of commodities ... must be sold. If this is not done, or only partly accomplished ... the laborer has been none the less exploited, but his exploitation does not realize as much for the capitalist ... The realization of surplus value ... is not determined ... by the absolute consuming power, but by the consuming power based on antagonistic conditions of distribution, which reduces the consumption of the great mass of the population to a variable minimum.

He also argued that there was a relationship between realization crises and the exploitation of workers:

> The epochs in which capitalist production exerts all its forces are always periods of overproduction, because the forces of production can never be utilized beyond the point at which surplus value can be not only produced but also realized; but the sale of commodities, the realization of the commodity capital and hence also of surplus value, is limited not only by the consumption requirements of society in general, but by the consumption requirements of a society in which the great majority are poor and must always remain poor. (Marx, 1885)

We could modify our neo-Marxian model by introducing aggregate demand considerations, perhaps by introducing a desired accumulation function as in the neo-Keynesian model of chapter 2 (see Harris, 1978).[30] We do not make this modification because although Marx discussed in detail the possibility of crises, he did not provide a rigorous analysis of their actual occurrence, leaving for Kalecki and Keynes to show how output and income could fall in response leakages from the circular flow; thus effective demand is more appropriately considered in the neo-Keynesian and Kalecki-Steindl models.

We conclude this section with a remark on an approach which is very similar to the neo-Marxian approach. This approach, suggested by Sraffa (1960) (see also Pivetti, 1985, Panico, 1985) determines r exogenously. The implications are the same as in the cases in which V or ϵ are fixed, in the absence of other changes, although the effects of technological changes (or other parametric variations) could obviously be different. Note also that, while this approach fixes r in terms of monetary factors, this does not imply that class struggle does not determine distribution: it can do so by affecting government policy which determines the interest rate and hence the rate of profit.

3.3 The neo-Keynesian approach

3.3.1 The neo-Keynesian approach and closure

Although Keynes (1936) was not directly concerned with long-run issues relating to growth, his approach has been extended to deal with questions of growth (and in some cases distribution) by Robinson, Kaldor, and Harrod, among others.

The central feature of this approach (which we call the neo-Keynesian one) is the stress on effective demand, which was absent in the neoclassical approach and our representation of the neo-Marxian approach. The neoclassical version, for example, assumes that full employment saving is always invested, either because saving is for investment, or because they are brought into equality by variations of the rate of interest. According to Keynes, if in an uncertain environment firms do not foresee good prospects for the future, they curtail their investment plans, and, with no automatic mechanism to equate investment to full employment saving, output and employment would have to fall, resulting in unemployment.

Although a critic of neoclassical theory (and perhaps because of that), Keynes used a fairly neoclassical approach; firms were price takers and profit maximizers and (except in isolated instances) consumers were identical. However, his emphasis on the aggregate consumption function reflected a departure from the neoclassical insistence on deriving behavioural relations from individual optimization. This approach was adopted also in his depiction of investment behaviour in an uncertain environment (despite his preliminary discussion in neoclassical terms), and his emphasis on group psychology and animal spirits.

Neo-Keynesians follow Keynes in using independent investment functions or firms' desired accumulation functions. Harrod (1939, 1948, 1973) assumed a function embodying the acceleration principle, of the type

$$I = \mu dX/dt, \tag{3.18}$$

where μ is a constant (which may equal a_1) and for simplicity it is assumed that actual and expected changes in output are equal. Equilibrium requires saving-investment equality times, so that, with saving given by (3.1),

$$\hat{X}^w = \sigma/\mu \tag{3.19}$$

which is Harrod's warranted rate of growth. Harrod showed that the economy is on a knife edge: if its actual rate of growth is different from the warranted rate, it will tend to move further away.[31] Suppose the actual rate of growth is lower than the warranted rate, so that $\hat{X} < \sigma/\mu$ (with a similar argument holding for the other case). This implies, from (3.1), (3.18), and (3.19), that $I < S$, so that there is an excess supply of goods. If excess supply induces firms to reduce the rate of change of output, the rate of growth of the economy will fall further.

The knife-edge instability property follows from the desired investment function assumed by Harrod, given his choice of the adjustment mechanism and the saving function. The equilibriating variable in his analysis is the rate of change of output, not the level of output within a short-run period.[32] But the rate of change in output here affects only investment, not saving (which depends on the *level* of output). As is well known, greater responsiveness of investment than of saving to the adjusting variable, inevitably results in instability.

Developments in business cycle theory accompanied those in growth theory. Samuelson (1939) and Hicks (1949, 1950) combined the multiplier with the accelerator, and, by introducing time lags in consumption and investment functions, demonstrated the possibility of cyclical behaviour.[33] While these models ignored the effect of investment on the stock of capital, others took it into account.[34] Thus Kaldor (1940) made investment and saving depend positively, but in a non-linear manner, on the level of activity (the dependence of investment on the level of activity made his model depart from the accelerator models which made investment depend on the change in the level of activity) and on the stock of capital (investment negatively and saving positively). Further, by assuming that the level of activity changed according to the difference between investment and saving, and the level of capital stock changed according to the rate of investment, he showed that cycles could result.[35] Goodwin (1951) assumed that investment depended on the difference between the desired and actual stocks of capital, the former being proportional to the level of output, and, combining this with the multiplier, showed that cycles occurred due to capital stock overshooting (and undershooting) its desired level. While these models do not allow for sustained increases in capital stock over time, Goodwin (1955) introduced growth by allowing a component of desired stock or capital to grow exogenously due to technological progress, very much along the lines of the contributions which will be examined in chapter 5. Since these models allow the level of output to adjust to the saving-investment gap, they can be interpreted as being more in the Kalecki-Steindl tradition than in the neo-Keynesian tradition in which output is at full capacity.[36]

Developments in what we may call neo-Keynesian growth theory proper modified both adjustment mechanisms and the nature of the desired accumulation function. Kaldor (1955–6, 1960) emphasized the role of income distribution as an adjusting variable: differential saving propensities between income classes imply that distributional shifts due to price changes (given money wages) adjust for imbalances between demand and supply, as was assumed in section 2.3.2. This forced saving mechanism was not emphasized by Keynes (1936), who allowed output (in addition to prices) to adjust in the short run to bring demand and supply to equality, and did not emphasize income distribution.[37] As for the desired accumulation function, while sometimes maintaining an accelerator mechanism or an output variable, the emphasis shifted to making the rate of accumulation depend positively on the expected rate of profit (Robinson, 1962). Adding the assumption that desired and expected rates of profit are equal, we obtain the investment function of our neo-Keynesian model.[38] This type of function is not to be found in Keynes (1936), where investment depended on the interest rate and exogenously given long-period expectations. But our model without assets cannot accommodate the interest rate (neo-Keynesians have underplayed its role in any case), and for long-run analysis it is reasonable to endogenize expectations by introducing the rate of profit.

The two leading participants in these developments appear to be Robinson and Kaldor.[39] Kaldor (1960), as stated above, assumed long-run full employment growth, so that its rate was determined by the growth rate of labour supply and technological dynamism. This full employment assumption puts him more in the neoclassical tradition,[40] and, since we are assuming given technology, the study of technological dynamism is postponed to chapter 5. Our model has thus followed Kaldor only by using his income distribution theory, the analysis of growth following Robinson (1956, 1962) instead.[41]

3.3.2 Modifications of the neo-Keynesian model

Since Keynes allowed for technical substitution and maximizing behaviour, and ignored class distinctions, it is of some interest to examine the generality of the basic neo-Keynesian model by incorporating these neoclassical elements into it, although retaining its distinctive features: the desired accumulation function and less than full employment growth.

If instead of fixed coefficients the production function is given by (3.5), and firms maximize profits in perfectly competitive markets, the marginal product of labour will equal the real wage satisfying (3.7) and, by the adding up theorem (even without credit markets or the renting of capital), (3.6). This implies, as in the neoclassical model, that the wage-profit frontier is convex to the origin. Maintaining all of our earlier neo-Keynesian assumptions, equilibrium g and r are determined exactly as they were in Figure 2.4, and the rest of the variables are determined in the same way; the only difference is that the wage-profit and consumption-growth frontiers will now be convex to the origin as in Figure 3.4. If we replace the class-specific consumption assumption (A.12) by (3.1), where σ is constant, we get $S/K = \sigma/a_1$; since a higher r implies a higher a_0 and a lower a_1, the saving (S/K) line becomes concave, so the diagram needs a minor modification.

In the short run, when K and N are given, with r determined to bring saving and investment to equality, equation (3.6) determines a_1/a_0, the capital-labour ratio. Since all K must be utilized, this determines L, and there is no reason for L to equal N. We assume, for unemployment to exist, that the equilibrium satisfies $L < N$. The short-run equilibrium is also the long-run equilibrium, except that K and L grow at the same rate.[42]

If we depict the representative firm, as in Figure 2.9, for this case, we get Figure 3.6. Given W and K, MC is the short-run marginal cost curve given by W/MPL where MPL is the marginal product of labour. Assuming (A.12) and that I is given, the AD curve is represented as before by

$$P = (sWa_0X)/(sX - I),\qquad(2.30)$$

but, since a_0 now increases with X, the curve will be sloped differently (and it could even have a positive slope). Competitive, profit maximizing producers will be price takers and choose output so as to set (expected) price equal to marginal cost. Given output, the AD curve will determine the (actual) price. Equilibrium will occur at the intersection of the AD and MC curves, with price P^* and output X^* (where expected and actual price are also equal). Note, unlike Figure 2.9, output need not be at the vertical part (which need not even exist) of MC.

A rise in the desired level of investment will push AD upwards, raise the price, given output, and induce a higher level of output as producers try to maximize profits at the higher price. If we take the general form of the desired accumulation function, this will shift the desired accumulation curve up, raise the rate of profit and the rate of growth, and reduce the real wage, implying a higher a_0 and a lower a_1. This interpretation of the Keynesian model, where price changes instantaneously, and output reacts to price changes, is more consistent with Keynes's own ideas than to Leijonhufvud's (1968) interpreta-

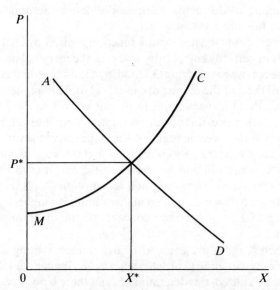

Figure 3.6   Neo-Keynesian model with variable coefficients: the short run

tion.[43] The adjustment here should also be contrasted to that in the fixed coefficients case, where, with the economy always at full capacity, quantity could not adjust in the short run.

3.4 The Kalecki-Steindl approach

3.4.1 The Kalecki-Steindl approach and closure

While the three previous approaches are widely discussed in the literature, the approach we have named after Kalecki and Steindl has attracted relatively little attention.[44] While Steindl has almost completely been ignored, Kalecki has not done much better. He has often been dismissed as someone who co-discovered (or in more sympathetic interpretations, discovered before and in a superior form) Keynes's theory of aggregate demand.[45] Recently, however, Kalecki's work has received more attention.[46]

Kalecki's (1954, 1971) approach departed much further from the neoclassical approach (than did Keynes's), it was firmly rooted in the classical/Marxian tradition, and shared some elements with Keynes. Rather than using the neoclassical assumption of individual optimizing behaviour in a perfectly competitive environment, he developed an alternative microtheory in which oligopolistic firms set their price by adding a markup to prime costs,[47] where the exogenously given markup rate depends on the 'degree of monopoly power' which incorporates several features including the degree of industrial concentration. This set him apart from Keynes who employed optimizing behaviour within a perfectly competitive framework.[48] His Marxian roots made him emphasize class distinctions: capitalists had a higher propensity to save than workers, and the forces relating to class struggle also affected the markup. But like Keynes he emphasized the role of aggregate demand by introducing a desired accumulation function. In his simpler models investment is autonomous, and savings adjust to it (so that output adjusts to demand). In fuller analyses, he argued that investment depends on the availability of finance, changes in profitability, the level of capital (a higher stock of capital reducing the need for investment),[49] and technological change (which does not concern us now). Note that Kalecki's analysis implies that the rate of profit can affect investment by affecting profit expectations (as in the neo-Keynesian approach), but also higher profits allow greater business savings, and thus (given credit market imperfections) ease the financial constraints to investment.

Steindl (1952) followed Kalecki in assuming that firms operate with excess capacity, but for him the utilization of productive capacity was fundamental to the investment planning of the firm. Confronted with a decline in demand, firms will cut back output (as in Kalecki) rather than prices, but the resulting fall in output will reduce capacity utilization, and therefore reduce investment and growth. This dependence of desired accumulation on capacity utilization makes us name the approach after both Kalecki and Steindl.[50] The rate of capacity utilization may affect investment for several reasons. First, firms may want to hold excess capacity due to fluctuations in demand, or expected growth in demand which, given indivisibilities in capital equipment, may make it profitable for them to build ahead of demand. When the utilization of capacity falls below the desired level, firms will want to increase the utilization and invest more slowly, and conversely when the utilization is above the desired level. Secondly, the level of output (as a ratio of capital stock) may signal to firms the strength of the market, exciting animal spirits as in accelerator models. Notice also that a higher capital stock, *ceteris paribus*, implies a lower utilization of capacity, and this may depress investment, along the lines described by Kalecki.[51]

Despite their shared emphasis on aggregate demand, there are two important differences between the neo-Keynesian and Kalecki-Steindl approaches, as incorporated in our models of chapter 2. First, the former assumes full capacity utilization (usually with perfect competition) and that price varies to clear the market, while in the latter firms follow markup pricing and operate with excess capacity and output varies to clear the market. Secondly, while in the former investment depends on the rate of profit, in the latter it depends both on the rate of profit and the rate of capacity utilization.

The main difference between the Kalecki-Steindl model and the others (as shown in chapter 2) is that it implies, for given technology, that a higher real wage is associated with higher rates of profit and growth, while the others imply inverse relations. This model thus formalizes the underconsumptionist arguments of Sismondi, Rodbertus, Hobson, Baran, and Sweezy, among others.[52] Since, as seen above, this result depends crucially on the existence of excess capacity at long-run equilibrium, it is important to examine this feature of the model more carefully.

It has been argued (Eatwell, 1983, Committeri, 1986, Auerbach and Skott, 1988) that the existence of excess capacity is inconsistent with long-run equilibrium. While Eatwell gives no reason for this claim, Committeri and Auerbach and Skott essentially argue that unless actual and desired (they call it 'normal' and 'optimal', respectively) rates of capacity utilization are equal firms will not be in (steady state) equilibrium. Committeri argues that if firms have excess capacity in this sense they will want to reduce investment, and this will lead to a reduction in aggregate demand and an increase in excess capacity, so that the economy will slip down a Harrodian knife edge. Auerbach and Skott argue that while equilibrium with excess capacity is possible for the short run, since the short-run investment response to changes in the utilization rate may be smaller than saving response, in the long run this inequality is reversed, so that this type of equilibrium cannot occur. Aside from the fact that their argument is based on an unsubstantiated *empirical* claim regarding the ranking of response parameters, which in any case cannot undermine the internal *logic* of the Kalecki-Steindl model, their assumption does not necessarily imply full capacity utilization in the long run, but only that the economy is balanced on the knife edge. The thrust of these criticisms is that since the model suggests that the economy is unstable, which does not seem to be confirmed by empirical evidence, there must be something wrong with the model and its conclusion of excess capacity at long-run equilibrium.

There is another way, however, to interpret the critics' conclusion, and that is, that the adjustment process considered by them, based on the distinction between the actual utilization and a given desired utilization rate, may be implausible. Our formal analysis of the model does invoke the idea of a desired rate of capacity utilization, although we have informally mentioned it.[53] Even if we assume that firms have such a desired rate, it may not be meaningful to think that it is unique even at long-run equilibrium. Recall (from chapter 1) that this equilibrium is not necessarily a tranquil state, but one in which the parameters of the model could be changing in ways unknown to the firms, and the economy may even be hovering around the equilibrium; thus firms may be in an uncertain environment. In such a situation, as Hicks (1979) noted, firms will want liquidity, which they obtain in addition to holding money, by holding excess capacity. More relevant for our purposes, the firms will not be able to optimize and find a unique desired rate of capacity utilization. As Boland (1985) argues:

> Since every decision takes time to implement, during that time the original givens ... might have changed and thus the implemented choice decision might not be the optimum for the new givens ... (T)he appropriate optimum (with regard to excess capacity or liquidity) may not be knowable by the firm since the knowledge of it depends on the unknown contemporaneous actions of other people as well as the unknown future.

It follows that, rather than introducing a desired capital-utilization ratio into our model, it is more meaningful to have an investment function which simply says that higher utilization rates, other things constant, raises investment. Note also that the utilization rate can enter the investment function for accelerator type reasons, which have nothing to do with the difference between actual and desired rates of capacity utilization, rendering these criticisms irrelevant.

A similar criticism may proceed by arguing that, given excess capacity, firms may try to increase their sales by reducing their markups, and thus take the economy to full capacity. While this is a possible consequence of excess capacity, Kalecki's writings lead us to doubt it: firms may raise rather than reduce markups to cover high fixed costs; the dynamics of industrial concentration may tend to maintain excess capacity (see below); and high unemployment usually associated with excess capacity may reduce the power of workers and raise the markup (see chapter 4).

3.4.2 Modifications of the Kalecki-Steindl model

The rest of this section will examine some modifications of the basic Kalecki-Steindl model and examine more fully its relations with the other approaches.

First we relax the fixed coefficients assumption. Although Kalecki assumed that average variable and marginal costs for industrial firms (for given wages) are constant at normal ranges of operation, and empirical investigations confirm this, it is worth examining whether the model's results can withstand the modification of the fixed coefficients assumption. It may be argued, for instance, that, if capital can be substituted for labour, firms with idle capacity will not hire labour.

Note that the possibility of technological substitution does not necessarily imply that factor substitution will actually occur. If the markup and the labour-output ratio are given, the markup equation implies a fixed real wage. The level of capital *used* by firms is not fixed in the short run, since they can leave capacity partially idle: thus given the real wage and constant returns to scale, cost-minimizing firms will choose to be on a ray through the origin on their isoquant map (up to full capacity), consistent with our assumption of a constant labour-output ratio. Thus the marginal cost curve of firms will be horizontal as in chapter 2, though it will be upward rising rather than vertical when 'full' capacity is reached and the labour-output ratio is a variable.[54] Given excess capacity our short-run analysis therefore requires no modification; and, for given technology and markup, neither does our long-run analysis.

Nevertheless, to examine its implications, we assume that firms actually substitute capital for labour. Assume that the relation between a_0 and a_1 is given by (3.5), which implies

$$a_0 = G(a_1), \tag{3.20}$$

where $G' < 0$ is the marginal rate of substitution between labour and capital, and $G'' > 0$ due to diminishing marginal returns to factors. We may take $1/a_1 = X/K$ to be an index of capacity utilization, comparing it to its (perhaps asymptotic) maximum or other reference value. With capacity utilization defined in this way, variable coefficients does not imply full capacity utilization, although it does imply that machines will not be 'idle' in the sense that they are not used at all.

As before, we assume that the firms practise markup pricing so that (2.13) holds with a given z. The firms' short-run average variable cost curve is upward rising instead of L-shaped as in Figure 2.11: given K, a rise in X implies a fall in a_1 and a rise in a_0, so that Wa_0 increases with X. This implies that supply price increases with output.[55] The real wage is given by

$$V = [1/(1+z)]/G(a_1) \tag{3.21}$$

which shows that a higher capacity utilization implies a lower a_1, hence a higher a_0 and a lower real wage.[56] Substituting $a_1 = K/X$ in (2.3) we get

$$1 = Va_0 + ra_1. \qquad (3.22)$$

Using (2.13), (3.21), and (3.22) we get

$$V = 1/(1+z)G(z/(1+z)r) \qquad (3.23)$$

which is the equation for the negatively sloped wage-profit frontier for a given z. It also follows from (2.13) and (3.22) that

$$V/r = (1/z)(a_1/a_0)$$

which implies that given z the wage-profit ratio rises with the capital-output ratio. Since by (3.22) the slope of the wage-profit frontier is $-a_1/a_0$, as we move down the frontier and V/r falls, its slope must fall in absolute value; the frontier (for a given z) is thus convex to the origin. Since the relationship between C and g is given by

$$1 = Ca_0 + ga_1 \qquad (3.24)$$

and, since (2.4) holds (so that a_0 and a_1 change in the same way when g changes as when r changes), the consumption-growth frontier has a shape similar to that of the wage-profit frontier. The determination of equilibrium can be shown in the same manner as in Figure 2.7, except that the dashed linear wage-profit and consumption-growth frontiers have to be replaced by convex-to-the-origin ones just described. The intersection of GG and sr curves determine g^* and r^* as before, but now the wage-profit frontier determines V^* after r^* has been determined, since a_0 is no longer fixed.

Consider now the effect of an increase in z, which had striking implications in the fixed coefficients case. As before, this will shift the GG curve down, reducng g^* and r^*, demonstrating that a rise in the degree of monopoly reduces the rate of growth. It can be shown that dV/dz has the same sign as

$$e_{KL}\{1 + g_2[(z/(1+z))(s-g_1) - g_2]\} - z,$$

where $e_{KL} = -G'a_0/a_1$, and g_1 and g_2 are the derivatives of the desired accumulation function with respect to r and u. Since $g_2 > 0$ and the expression within square brackets must be positive for the short-run stability of the model, it follows that, if the elasticity of substitution in production is large enough for $e_{KL} > z$, the real wage will rise with z, so that there will be no inverse relation between the growth rate and the real wage. This happens because when the markup increases a_1, due to a reduction in aggregate demand, there is a decrease in a_0 which exerts an upward pressure on the real wage. In the fixed coefficients case $e_{KL} = 0$ and the real wage fell when z increased. A low elasticity which makes the above expression negative can continue to produce this result in the variable coefficients case. Thus substitution in production may, but does not necessarily, destroy the positive relation between the wage rate and the rate of growth (although since z is the profit share there is still a positive relation between the wage *share* and the growth rate).

Secondly, we examine the role of optimization in the theory. We confine attention to the markup rule; explanation of investment and saving behaviour in terms of optimization is also possible, but is not analysed here. The markup rule may be criticized as being inconsistent with optimizing behaviour, and may therefore be unattractive to those trained in a neoclassical perspective. There are two possible responses to this: one is to argue that markup pricing can be derived from optimizing principles, and the other to argue against their use.

Regarding the first, recall that the model examines firms in an imperfectly competitive environment. Suppose, for simplicity, that the economy has many firms (so that each can ignore the reactions of others), each producing a different good, acting as a monopolist in its own market. Assume that all firms face identical cost and market conditions,[57] and that they have a given perception of the elasticity of demand, e_d, of their product. With labour as the only variable factor and the wage given at W, profit maximization implies

$$P = [e_d/(e_d - 1)]\ W/MPL,$$

where MPL, as before, is the marginal product of labour. Assuming that MPL does not change with output and is equal to average product of labour, and writing $z = 1/(e_d - 1)$ we get the markup pricing equation (2.13).[58] Two caveats are in order. First, this identification is not possible with a variable MPL. Secondly, this interpretation is narrower than Kalecki's: while greater competition (more firms) can increase e_d and lower z as in Kalecki's theory also considers other aspects of concentration, such as interfirm effects,[59] as well as other factors such as union power.

Regarding the second, note that firms generally operate in an uncertain environment, information relating to cost is 'hard', while that relating to demand, which cannot be reduced to a hard status by actuarial expected-value calculations, is 'soft'. Firms can then be thought of as making pricing decisions by using only the former, following the rule of thumb of setting the price by adding a markup on prime costs.[60] The neoclassical optimizing procedure can thus be criticized for ignoring distinctions between different types of information.

Thirdly, we compare the neo-Keynesian and Kalecki-Steindl approaches in light of some modifications suggested now and in section 3.3.2. We earlier emphasized the fact that, in the former approach, changes in demand are accommodated by changes in the price with output at full capacity, while in the latter the accommodation was through changes in capacity utilization, with the price level fixed. With variable coefficients this distinction seems to disappear, since changes in demand lead to price and output changes in both models. However, in the neo-Keynesian approach with price-taking firms, a rise in demand raises the price and profit-maximizing firms then change output levels, while in the Kalecki-Steindl approach firms respond by changing output, which affects the price by changing prime costs.

This distinction between the two approaches, however, can be overemphasized. If we assume that Kaleckian firms have a supply price determined by the markup and prime costs, and change levels of output in response to deviations in the demand price from the supply price (as in section 2.3.4), the two models seem to share the property that the market adjusts the price and firms respond by adjusting output. The only difference between them (aside from investment functions) is that one (usually) assumes perfect competition and the other does not.

To bring them closer we can modify the neo-Keynesian model by assuming that (identical) firms perceive a falling isoelastic demand curve for their product. In this case, as discussed above, profit-maximizing firms will set price according to

$$P = (1 + z)W/MPL$$

(where z depends on the elasticity of demand) which, given the production function (3.5) implies

$$V = 1/(1 + z) \left[f(a_1/a_0) - (a_1/a_0)f' \right]. \tag{3.25}$$

This shows that a higher a_1/a_0 is associated with a higher V. Equations (3.22) and (3.25) imply

$$r = 1/a_1 - 1/(1 + z) \left[f(a_1/a_0)/(a_1/a_0) - f' \right]. \tag{3.26}$$

Since the term within square brackets (the difference between the average and marginal values of the function (3.5)) and a_1 rise with a_1/a_0, (3.26) implies that a higher a_1/a_0 is associated with a lower r. All this implies that a_1/a_0 falls when V/r falls, so that the wage-profit frontier for the given z is convex to the origin as in the variable coefficients Kalecki-Steindl model.

If we assume that the rest of the model is the standard neo-Keynesian one, so that equations (2.4) and (2.10) hold, g^* and r^* are determined exactly as in Figure 2.5. In this case a rise in z does not change saving and investment curves, leaving g^* and r^* unchanged; equation (3.26) shows that a_1/a_0 will rise and (3.25) shows that V^* will fall in consequence. If we replace (2.10) by (2.14), so that desired accumulation depends also on the rate of capacity utilization, then the distinction between the neo-Keynesian and Kalecki-Steindl models disappears altogether. We see that (3.26) implies a positive relation between $1/a_1$ and r, given z, which can be substituted into (2.14) noting that $1/a_1 = X/K$, to yield the GG curve given z. The GG and sr curves determine g^* and r^* as in Figure 2.7. An increase in z, from (3.26), implies a higher r for a given a_1; to keep r the same this requires a higher a_1, which, from (2.14), implies that the GG function shifts down. Thus g^* and r^* fall, and there is also possibly a fall in V^*.

It thus appears that the crucial difference between the two approaches results from the specification of the investment function, and, if we use the same one, given by (2.14), the two models have similar implications. Given this similarity, it is tempting to argue that there is no need to distinguish between the two approaches. However, the modified model considered here, with its emphasis on changes in capacity utilization and its (possible) positive relationship between growth and distribution, behaves more like the basic Kalecki-Steindl model than the basic neo-Keynesian one. Had we combined them under the better-known neo-Keynesian banner, it is likely that these insights obtained from the Kalecki-Steindl model would have been lost. This has been our justification for treating them as alternatives, to distinguish between demand-constrained economies which are close to full capacity and those with generalized excess capacity.

Fourthly, we examine the relationship of the model to the Marxian approach. Note first that the Kalecki-Steindl approach (in its fixed coefficients form) and the Marxian one fix the real wage exogenously: the latter, by subsistence or class-struggle considerations and the former, by fixing the markup rate, reflecting the degree of monopoly power. Further, recalling the definition of the rate of exploitation, $\epsilon = (1 - Va_0)/Va_0$, we see from (2.19) that

$$\epsilon = z,$$

so that Marx's rate of exploitation is Kalecki's degree of monopoly power. Taking the version of the neo-Marxian model which takes ϵ to be exogenous, it follows that the two approaches fix distribution in the same way, and they differ only on what factors this distribution parameter depends on. Marx seems to emphasize class-struggle forces while Kalecki the degree of monopoly, which includes some forces relating to class struggle.[61]

Despite these similarities, while the neo-Marxian approach ignores demand problems and assumes full capacity utilization, the Kalecki-Steindl approach emphasizes the role of demand.[62] This difference results in remarkable differences in the implications of distributional changes. A rise in ϵ in the neo-Marxian model reduces the real wage, raises the rate of profit and the rate of extraction of surplus value, increasing saving, investment, and growth. In the Kalecki-Steindl model a higher ϵ reduces the real wage, but is not able to increase the rates of profit and growth, because the lower real wage reduces the demand, resulting in problems of realization ignored in the neo-Marxian approach. This is not to say that Marx did not discuss the problem of demand: indeed, as our discussion of Marx's analysis of the realization problem suggests, Marx's own approach is closer to the Kalecki-Steindl model than is the neo-Marxian model.[63]

Finally, we modify the model to endogenize the markup rate. To examine how it can change due to secular changes in industrial structure,[64] we consider a function which relates its time-derivative, dz/dt, to g and z,

$$dz/dt = Z(g, z). \tag{3.27}$$

The derivative of Z with respect to g, which will be taken to be an indicator of the rate of growth of industry,[65] can be assumed to be negative for the following reasons. Since with fast growth new entrants are encouraged to enter industries through the attraction of higher profits, and barriers to entry appear less formidable with expanding markets, we may expect an inverse relation between g and changes in ratios of industrial concentration. Since collusion seems effective the greater the share of larger firms, we may expect a positive relation between concentration rates and markups; a large empirical literature surveyed in Weiss (1980) and Scherer (1980) confirms this. Given a (roughly) linear relationship (as is generally estimated) we can obtain a positive relationship between dz/dt and changes in concentration rates. Combining these two relationships implies the negative sign of the derivative.

The derivative with respect to z can be assumed to be positive for lower levels of z and negative for higher levels. At low levels, an increase in z, reflecting a rise in monopoly power, will increase the ability of firms to raise their monopoly power and markups (through the greater concentration of credit, for instance). But, beyond a certain level of z, further increases will reduce dz/dt because higher markups will induce greater entry and faster falls in concentration ratios (as suggested by limit pricing models of oligopoly), because existing firms may apprehend government action if they push up their rates of markup excessively, and because firms cannot push up the markups indefinitely in any case.

Equation (3.27), given these assumptions, yields a relation between g and z, with z stationary, which can be drawn as the $dz/dt = 0$ curve of Figure 3.7. As shown by the arrows, z falls above the curve and rises below it. The position of this curve can be assumed to depend on the structure of government policies: for example, government action promoting competition would shift the curve down to the position shown by the dotted line. We also draw the IS curve showing the (goods market) equilibrium levels of g for different levels of z; the curve is downward sloping for reasons discussed in section 2.3.4. While there are a variety of possible configurations of these curves, the figure shows one interesting possibility. Assuming that the goods market is always in equilibrium, the economy will always be on the IS curve. The movement of the economy in the very long run, when z moves, is shown by the arrows on the IS curve. A and B represent the two equilibrium points: A is unstable and B is stable. If the initial value of z is such that $z < z^*$, the economy will grow with rising g till it reaches full capacity utilization, at which time the economy will become neo-Keynesian (in its fixed coefficients version), but an economy starting from $z > z^*$ will over time tend to move towards B; if it moves down the IS curve it can be said to be stagnating due to the intersection of the forces of monopoly power.[66]

The markup may depend on short-run factors as well, but the nature of these effects is not obvious. During periods of high demand, firms may raise markups to take advantage of buoyant markets, and during periods of low

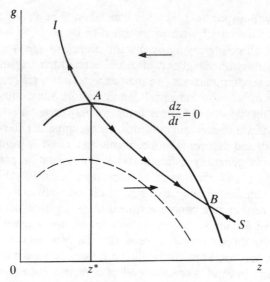

Figure 3.7   Endogenous changes in the markup in the Kalecki-Steindl
model

demand, they may reduce markups to increase sales. This seems to be a likely possibility when demand is very low, and some evidence (Scherer, 1980) has been produced to show that the markup falls during slumps. Moreover, it can be argued (following Eichner, 1976, Harcourt and Kenyon, 1973, Wood, 1975) that during expansions firms will want to invest more by generating higher internal saving, and therefore attempt to raise their markups (though in the Kalecki-Steindl model the actual outcome could be the opposite of what is intended, if firms do not sufficiently increase their investment). Despite arguments of this type, several reasons can be adduced to show that markups may fall with output.[67] First, Kalecki (1954) argues that the markup depends partly on the level of overheads relative to prime costs, and, since during recessions this ratio is high, the markup may rise. Secondly, Harrod's (1936) – admittedly questionable – analysis shows that during the upswing the elasticity of demand falls, so that firms would be charging a lower profit maximizing markup. Thirdly, neoclassical oligopoly theory provides some additional reasons. Stiglitz (1984), using an incomplete information model, argues that prices may be lower since limit pricing may be more prevalent during booms if the threat of potential entry is greater at that time. Rotemberg and Saloner (1986) argue that oligopolies can find implicit collusion more difficult when demand is relatively high, since the benefit to firms of undercutting price and capturing a larger share of the market is likely to be higher than the possible loss from being punished later, so that they will decide to set lower prices to prevent undercutting. They also empirically question the proposition that markups tend to be lower during slumps. Goldstein (1985) uses a neo-Keynesian (featuring a desired accumulation function) price leadership model to show that, in the face of foreign competition, the optimizing price leader will increase the markup in the early expansion, but reduce it later in the expansion.

An interesting implication of the short-run variability of the markup is that it could imply cycles. A simple story using some of the arguments just given could be that a high level of demand causes the markup to rise, which redistributes income to profits, reduces economic activity and growth. This eventually creates a slump, reduces the markup, and results in a bout of expansion.[68] It is obvious from the arguments given above that this is not the only possible story, and in fact some could involve a cumulative expansion or collapse.

3.5 Conclusion

This chapter sought to justify the use of the labels used for the four basic models of the previous chapter, and to extend them in several ways. It has shown that the introduction of organizing principles such as the optimizing approach of neoclassical economics and the class struggle approach of Marxian economics does not alter the substance of the models. It has also shown that the introduction of several other modifications, including variable coefficients, does not alter the general conclusions of the previous chapter, although the models become more complicated and to some extent the distinction between the neo-Keynesian and Kalecki-Steindl approaches becomes blurred.

Though this chapter has made some modifications of the basic models, it has not introduced money and inflation, technological change, and multi-sector complications. These issues are examined in the next three chapters, but − for simplicity − by ignoring the issues raised in this one.

CHAPTER 4

Money and inflation

4.0 Introduction

While we have so far considered only 'real' economies which do not explicitly contain money (or other assets),[1] in several of the traditions we have discussed, particularly the neo-Marxian and neo-Keynesian ones, money plays a central role. This chapter introduces money into the models and examines the determination of the monetary variables, or rather their rates of growth, the most important being the rate of inflation.

The four sections of this chapter introduce money and inflation into our four basic models, sometimes developing hybrids which introduce elements from different approaches. Our purpose is to provide a flavour of how money can be incorporated into the models and how inflation can be explained in them, and not to provide an exhaustive treatment of these issues in the alternative approaches. We thus introduce money into the models without detailed treatments of asset markets, monetary and credit systems, and the nature of money creation,[2] simply using the quantity equation of the form

$$M\tau = PX,$$

where, in addition to the variables introduced in earlier chapters, M is the total quantity of money, and τ is the (income) velocity of circulation. Logarithmic differentiation yields

$$\hat{M} + \hat{\tau} = \hat{P} + \hat{X}, \tag{4.1}$$

where the hats over the variables denote their rates of growth. Our task will be to examine how the new variables are determined in the different models.

4.1 Neoclassical models

4.1.1 A simple neoclassical monetary growth model

The neoclassical approach emphasizes the role of money as a medium of exchange and defines it to include all instruments serving such a function. Its rate of growth is assumed to be fixed by the monetary authorities, and τ is assumed to be a constant determined by factors such as payment practices and the technology of money circulation, so that $\hat{\tau} = 0$. Thus (4.1) implies

$$\hat{M} = \hat{P} + \hat{X}, \tag{4.2}$$

where $\hat{M}$ is exogenous.

The simplest way to incorporate this into the neoclassical model of section 2.2.1 is to append this equation to the rest of the model, given by (2.1), and (2.3) through (2.6). The real variables are solved as before, and noting that, with full capacity utilization, $\hat{X} = g$, we get

$$\hat{M} = \hat{P} + g \tag{4.3}$$

which solves for $\hat{P}$. Note that an (exogenous) rise in $\hat{M}$ will raise $\hat{P}$ by the same amount, leaving unchanged the real variables. This establishes the super-neutrality of money, and the dichotomy between the real and monetary sectors. The result supports the Monetarist claim that high rates of inflation are due only to high rates of growth of money supply so that $\hat{M}$ should be set roughly equal to $\hat{X}$ for price stability. These results can easily be shown to carry over to the case of variable coefficients of production.

4.1.2 Some comments and modifications

We now make several comments on this model, which lead to some simple modifications which bring it closer to the more standard neoclassical monetary growth models.

First, there is the well-known objection that (4.3) lacks a mechanism: what causes $\hat{P}$ to rise when $\hat{M}$ rises? The mechanism usually suggested is the so called 'real balance' effect (Pigou, 1943, Patinkin, 1965) by which an excess supply of money leads individuals to buy more goods, driving up the price. While such a mechanism is not present in our model, it can be introduced in a simple manner. Assume that the saving rate of capitalists, s, depends on the excess of the rate of growth of money supply over the rate of growth of money demand. The higher this excess, the lower is s, with capitalists spending more on consumption (workers' consumption, being equal to income, is not changed). We thus write

$$s = s(\hat{M} - \hat{P} - \hat{X}), \tag{4.4}$$

where $s' < 0$ and with $1 > s > 0$ for all relevant values of the argument.[3] The resulting model, however, has the same long-run equilibrium properties as the earlier one. This is obvious from the fact that in equilibrium (4.2) must hold, so that any change in $\hat{M}$ will result in just the change in $\hat{P}$ which leaves s unchanged at $s(0)$.

Secondly, it may be argued that the assumption that $\hat{M}$ is given exogenously, as determined by the monetary authorities, is inadequate in an economy in which the means of payment consist not only of currency but also of inside money. In this case, if we assume that all inside money consists of bank deposits, it can be shown that

$$M = [(cu + 1)/(cu + re)]H, \tag{4.5}$$

where cu is the currency-deposit ratio chosen by capitalists, re is the reserve-deposit ratio chosen by firms, and H is the stock of high powered money.[4] If we now assume that cu and re are fixed, and the monetary authorities fix $\hat{H}$, $\hat{M}$ gets fixed as well. Given these assumptions, we may assume that, even with inside money, $\hat{M}$ is fixed by the monetary authorities.

Thirdly, a problem with the model is that it relies on the traditional quantity equation which focuses only on the role of money as a medium of exchange, and ignores the role of money as a store of value. Once money is considered as an asset, there arises the issue of choice between different assets. Consider, for example, following Tobin (1965), that physical capital and money are the only assets. The rental of capital, r, is the return on holding capital, and the return on holding money, in real terms, is $-\hat{P}$; the yield differential is therefore $r + \hat{P}$. Assume that asset holders allocate their portfolios so as to demand capital and money in a ratio k depending on the yield differential, so that

$$k = k(r + p) \quad k' > 0.$$

In equilibrium, the assets must be demanded in the same ratio as they are supplied, so that

$$PK/M = k(r + p). \tag{4.6}$$

The rate of growth of the left-hand side is $\hat{P} + g - \hat{M}$. Since in equilibrium r and p are constant, the right-hand side of the equation is also constant, so that in equilibrium it implies (4.3). Thus this modified model has the same long-run equilibrium as the model of section 4.1.1, and has the same implications.

Finally, an objection may be made that the treatment of money in the model is essentially flawed, and money should and can be introduced into neoclassical models to play a real role. Neoclassical models allowing money to play such a role have been presented by Tobin (1955, 1965), Stein (1969), and others. This can be done most simply by using (4.6) and replacing (4.4) by

$$s = s(PK/M)$$

with $s' > 0$; in this case a change in $\hat{M}$ can be shown to affect r^*.[5] If variable coefficients of production are also allowed, the long-run equilibrium capital-labour ratio will also be affected (although not g which is still fixed by n).[6]

4.2 Neo-Marxian models

The neo-Marxian model can be extended to incorporate money and inflation in the same way as we did for the neoclassical model, that is, by appending (4.1) to it. In this case, as in the neoclassical model, the real variables will be solved in the real part of the model, examined in section 2.2.2, and with $\hat{M}$ exogenously fixed by the monetary authorities, and τ constant, (4.3) will determine $\hat{P}$. Thus, as in the neoclassical model, there is a dichotomy between the real and monetary parts of the model, and an increase in $\hat{M}$ merely raises $\hat{P}$ by the same amount.

4.2.1 Money and inflation in Marx

Although there is no reason to believe that the above model is not internally consistent, it is not the one that can be gleaned from Marx's own writings on money.[7]

Marx's discussion in the first part of *Capital* mainly examines gold, or commodity money. Even though he uses the quantity equation, the direction of causality is quite different. For Marx gold is the numeraire, and its price is determined, like that of any commodity, by its production price.[8] Without extending our one sector model to consider gold as another commodity, let us simply assume that the price level is determined by technology in gold production: if technological conditions in gold production do not change, $\hat{P}=0$. With $\hat{P}$ fixed, and with g fixed by the real side of the model, (4.3) determines $\hat{M}$. While the rate of growth of the amount of gold depends on the production of gold, all gold is not money. Gold not needed as money is hoarded, or used in some other form, and does not circulate as money. Thus, $\hat{M}$ adjusts to the level of transactions in the economy, given a constant V.[9]

While the neoclassical and Marxian theories both use the quantity equation and imply that the real part of the system determines the values of the real variables, their differences are fundamental. A change in the supply of money in the former model results in a one for one increase in prices (presumably by affecting the demand for goods), while in the latter money supply changes endogenously without causing inflation,[10] and the amount of gold beyond the needs of trade is hoarded. Money thus plays a more passive role in the Marxian system than in the neoclassical one.[11]

There are several questionable features in this neo-Marxian model. First, questions arise about hoarding. It plays a purely passive role in the model and there is no analysis of why wealth-holders want to hoard. Further, the model assumes that the amount of gold in existence is sufficient to meet the need for it. Secondly, the application of the theory to economies in which paper and credit money exist (so that all money is not gold) is problematic. For such economies, the price level (or its rate of growth) cannot be determined from the conditions of production of commodities; further the hoarding of paper currency in a passive fashion stretches the imagination more than does that of gold, which at least has some alternative uses. However, the existence of non-gold money *per se* does not create a problem. If paper money is convertible into gold (as in a gold standard system), paper money will circulate just like gold, and can also be hoarded in the same way. There can be a problem if too much paper money comes into existence, so that idle balances pile up, and individuals will try to exchange the paper for gold: a market discount of the paper against gold can spring up in the absence of convertibility at a guaranteed rate of exchange. This is not the same as the neoclassical quantity theory, since it only implies a specific decline in the value of paper money to commodity money in the market.[12] However, for economies not on the gold standard, there arises a need for an alternative theory of inflation and hoarding.

The essence of Marx's approach in interpreting (4.3) is to have $\hat{P}$ and g determined elsewhere in the system, and for $\hat{M}$ to be determined by it. In an economy with non-gold money not on the gold standard, several alternative mechanisms can support this approach. One possibility is to follow Foley (1983) in assuming that the value of money 'is linked closely to the dynamics of production and accumulation in a capitalist system, and to the factors which produce booms and crises'. More generally, the rate of inflation could be taken to be related to conditions in the goods and labour markets, rather than the conditions in gold production. In addition to hoarding and dishoarding, variations in $\hat{M}$ in a credit-money economy can be caused by the expansion and contraction of credit in response to changes in the demand for it.[13] A further possibility is to let the state vary $\hat{M}$ endogenously. Although this is in line with the Marxist concept of an endogenous state, one would have to explain why the state passively adjusts $\hat{M}$ to satisfy (4.3).

4.2.2 A simple neo-Marxian model of inflation

While other models which have neo-Marxian features will be considered later, here we examine a simple model, which is a hybrid of the neo-Marxian and neo-Keynesian models, as one interpretation of the approach just described. It is given by (2.1), (2.3), (2.4), (2.5), and (2.8), the five equations of the real model of section 2.22, (4.3), and

$$\hat{P} = \Omega[g^d(r) - sr], \tag{4.7}$$

where g^d is the desired accumulation function of the neo-Keynesian model,[14] and $\Omega > 0$ is a speed of adjustment constant.[15]

The equations of the real model solve for real variables, including r^*. Insertion of r^* in (4.7) solves for $\hat{P}^*$. We need to assume that the desired accumulation function is such that $g^d(r^*) > sr^*$ to ensure $\hat{P}^* > 0$. Note that in equilibrium firms are unable to invest at their desired rate, but their desire to do so causes inflation. After $\hat{P}$ is solved, equation (4.3) solves for $\hat{M}^*$. Figure 4.1 shows how equilibrium is determined. V^1 would have to be the real wage to allow firms to fulfil their investment plans, but, since class struggle puts the real wage at $\bar{V}$, there is an excess demand gap of $g^d(r^*) - sr^*$.

While this model – which is identical to Robinson's (1956, 1962) 'inflation barrier' model – is a hybrid with neo-Marxian and neo-Keynesian character-istics, neo-Marxian features dominate its real side (which is why we call it a neo-Marxian model) and the neo-Keynesian desired accumulation function

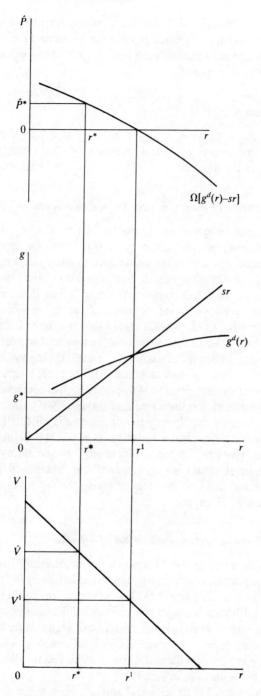

Figure 4.1   A hybrid neo-Marxian–neo-Keynesian model

only determines the rate of inflation. More buoyant animal spirits will only accelerate inflation in this model, but greater exploitation – reducing $\tilde{V}$ – will increase the rate of growth and reduce inflation. Thus an upward movement in $\tilde{V}$ can explain stagflation in this model.

4.3 Neo-Keynesian models

Neo-Keynesian models have to give a larger real role to the desired accumulation function. We examine a series of models which represent some well-known contributions to the literature on Keynesian models of inflation.[16]

4.3.1 The endogeneity of money in neo-Keynesian models

All the models we will examine here will use (4.1) where $\hat{M} + \hat{t}$ is assumed to be endogenous. Several reasons can be given to justify this assumption. Economists in the Keynesian tradition often argue that monetary authorities play a passive role. Weintraub (1978) argues that responsible political leaders will not tolerate large lapses from full employment, and writes that central banks merely ensure ample supplies of money to remove financial impediments to full employment and growth. Kaldor (1982) argues that the Central Bank's primary responsibility is to guarantee the solvency of the financial sector, and, since it cannot afford the consequences of a collapse of the banking system, it will change the supply of money to meet demand. Moore (1988) argues in favour of this position and adds that, since banks operate with excess reserves, and bank deposits are demand-determined (see also Lavoie, 1986), the credit system would also cause money supply to be endogenous. If all this is not enough and $\hat{M}$ does not adjust to satisfy our equation, as Kaldor (1982) argues, $\hat{t}$ can adjust to do the job. This can happen, for instance, if the economy is in a liquidity trap (in a general sense), so that money can enter and leave speculative balances without affecting the interest rate; speculators in effect will stabilize interest rates by changing $\hat{t}$.[17]

4.3.2 A neo-Keynesian demand-pull model of inflation

The first model contains the same excess demand theory of inflation that was examined at the end of the previous section, represented as before by (4.7). For the rest, we use (2.1), (2.3), (2.4), (2.5), and (2.6). The last equation could imply full employment growth (although it is also consistent with the assumption of a constant unemployment rate). While this assumption can arguably make the model un-Keynesian, we use it here since models with full employment and excess demand inflation have been built, and dubbed to be Keynes-Wicksell models, by Stein (1969, 1971) and Hahn (1969).[18]

In this model, which is a hybrid of neoclassical and neo-Keynesian models, the neoclassical features solve for the real variables. Figure 4.2 shows the

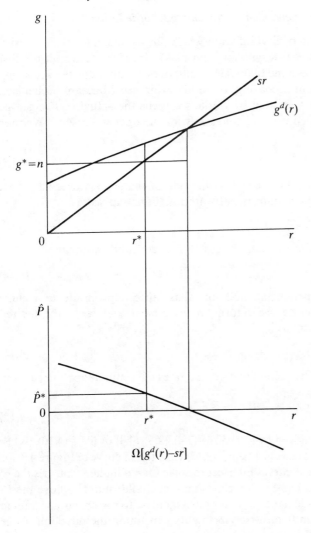

Figure 4.2 A hybrid neoclassical–neo-Keynesian model with full employment

determination of equilibrium. An upward shift in the desired accumulation function results in a higher $\hat{P}*$, while a rise in n increases $g*$ and reduces $\hat{P}*$.

Both this model and the one in section 4.2.2 contain the neo-Keynesian desired accumulation function, but their 'real' characteristics are determined by neo-Marxian and neoclassical elements. Further, they explain inflation in terms of what used to be called 'demand pull' elements.

4.3.3 A neo-Keynesian cost-push model of inflation

We now examine a model of inflation which allows unemployment to exist, makes the real part neo-Keynesian, and makes inflation the result of 'cost push' elements. We assume that wages and prices are not adjusted according to excess demand, but according to the following neo-Marxian mechanism: when the real wage $\tilde{V}$ desired by workers exceeds the actual real wage, the money wage is pushed up (and conversely for the opposite case).[19] We may formalize this with

$$\hat{W} = \Omega[\tilde{V} - V].$$ (4.8)

We may add a term which says that the rate of change in wages also rises, though with incomplete adjustment, with inflation, and write

$$\hat{W} = \Omega_1[\tilde{V} - V] + \Omega_2 \hat{P},$$ (4.9)

with $0 < \Omega_2 < 1$.[20] Since in equilibrium W/P is constant, we write

$$\hat{W} = \hat{P}.$$ (4.10)

In an economy experiencing inflation, it is more appropriate to include a capital gains term to the return on investment and write the desired accumulation function as

$$g^d = g(r + \hat{P}).$$ (4.11)

Finally, we assume that actual and desired rates of accumulation are equal in equilibrium, so that

$$g = g^d.$$ (4.12)

The model is thus given by (2.1), (2.3), (2.4), (2.5), (4.8) or (4.9), (4.10), (4.11), and (4.12). Equations (2.4), (4.11), and (4.12) yield the *IS* curve of Figure 4.3: for a given $\hat{P}$, the g^d and *sr* curves intersect to solve for r; a higher $\hat{P}$ pushes the g^d curve up, implying a higher r for goods market equilibrium.[21] Above the *IS* curve there is excess demand, so that r must increase to restore equilibrium in the goods market (and conversely below it), explaining the direction of the horizontal arrows. Equations (2.3), (4.9), and (4.10) give

$$\hat{P} = [\Omega_1/(1 - \Omega_2)][\tilde{V} - (1 - a_1 r)/a_0]$$ (4.13)

which also implies an upward rising relationship between $\hat{P}$ and r which we call the *PP* curve. If we assume that we are on the *PP* curve and not on the *IS* curve (say due to a movement of the *IS* curve), the figure shows that the return to equilibrium will be stable if the *PP* curve is flatter than the *IS* curve; we assume this to be the case.[22] The intersection of the two curves solves for the equilibrium values, r^* and $\hat{P}^*$. The lower part of the figure then solves for g^*.

An increase in the desired rate of accumulation shifts the *IS* curve down, implying a higher $\hat{P}^*$, g^*, and r^*. A rise in s shifts the *IS* curve up, reducing $\hat{P}^*$

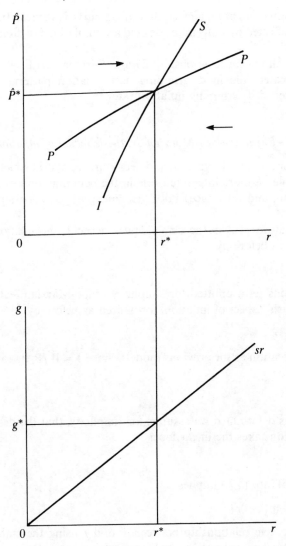

Figure 4.3   A hybrid neo-Keynesian–neo-Marxian model with 'cost-push' inflation

and r^*; although the sr curve will be rotated upwards, (4.11) and (4.12) show that it will also reduce g^*. A rise in $\tilde{V}, \Omega_1$ or Ω_2, will shift the PP curve upwards, also implying a higher $\hat{P}^*$, r^*, and g^* (and paradoxically, a lower V^*). Thus a strengthening in the position of workers, while fuelling inflation, also raises the rate of growth of the economy. Conversely, a weakening of the position of workers will reduce $\hat{P}^*$ as well as g^* and r^*. These results regarding the effect of class struggle on the growth rate disappear if $\hat{P}$ does not enter the desired accumulation function as an argument, so that the *IS* curve is vertical. In this case g^* will only be affected by shifts in the desired accumulation function and s.

This hybrid only allows for 'cost push' inflation. Further, neo-Keynesian features have a greater role in determining accumulation patterns, and neo-Marxian features in determining inflation rates.[23]

4.3.4 A neo-Keynesian-neo-Marxian synthesis model of inflation

We now consider another neo-Keynesian and neo-Marxian hybrid which lets the two traditions play a more integrated role in the determination of both growth and inflation, and introduces both 'demand push' and 'cost push' aspects of inflation.[24]

Neo-Keynesian features provide the 'demand push' aspect to the inflationary process, formalized as before by

$$\hat{P} = \Omega_1 [g^d(r) - sr] \tag{4.7'}$$

(with the capital gains term omitted for simplicity). Neo-Marxian features provide the 'cost push' aspect of inflation, formalized as before by[25]

$$\hat{W} = \Omega_2 [\tilde{V} - V] \tag{4.8'}$$

with the same justification as for previous models. Since $V = W/P$, its rate of growth, $\hat{V}$, is given by

$$\hat{V} = \hat{W} - \hat{P}. \tag{4.14}$$

To show how $\hat{V}$ is determined we assume, for simplicity, that the desired accumulation function takes the linear form

$$g^d = g_0 + g_1 r. \tag{4.15}$$

Substitution of (4.15) into (4.7') implies

$$\hat{P} = \Omega_1 [g_0 + (g_1 - s)r] \tag{4.16}$$

which implies a negative relationship between $\hat{P}$ and r using the stability assumption $g_1 < s$. It is depicted by line $\hat{P}$ in Figure 4.4; its horizontal intercept is given by $g_0/(s - g_1)$. Substitution of (2.3) and (2.5) in (4.8') implies

$$\hat{W} = \Omega_2 [\tilde{V} - (1 - a_1 r)/a_0] \tag{4.17}$$

which implies a positive relationship between $\hat{W}$ and r, depicted by line $\hat{W}$ in Figure 4.4; its horizontal intercept is given by $(1 - a_0 \tilde{V})/a_1$. The vertical distance between the lines $\hat{W}$ and $\hat{P}$ is $\hat{V}$, which is depicted by the upward rising line $\hat{V}$.

In long-run equilibrium, V must be stationary, so that $\hat{V} = 0$. Thus long-run equilibrium r in the figure is determined where $\hat{V} = 0$, which is given by

$$r^* = [\Omega_1 g_0 + \Omega_2 (1/a_1 - V^*)]/[\Omega_1(s - g_1) + \Omega_2 a_1/a_0].$$

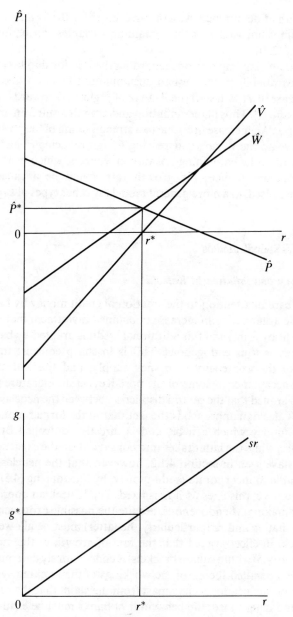

Figure 4.4 A hybrid neo-Keynesian–neo-Marxian model with 'cost-push'
and 'demand-pull' inflation

Notice that this occurs where $\hat{W} = \hat{P}$, which must of course be true due to
(4.14). We have drawn the curves so that $\hat{W} = \hat{P} > 0$ at long-run equilibrium;
for this to be true we must assume that $g_0/(s-g_1) > (1-a_0\tilde{V})/a_1$. To obtain
a long-run equilibrium in which there is a positive rate of inflation we make

this assumption. With r^* determined, we can solve for g^* for the lower part of the figure. The equilibrium values of the remaining variables can be found from (2.1), (2.3), and (2.5).

The model has interesting implications regarding the relationship between g^* and $\hat{P}^*$. An upward shift in the desired accumulation function due, for instance, by an increase in g_0, will shift the $\hat{P}$ curve of Figure 4.4 upwards. This will increase $\hat{P}^*$, r^*, and g^*. Thus higher inflation and growth result from more buoyant animal spirits. An increase in $\bar{V}$, due to a strengthening of the position of workers, will move up the $\hat{W}$ curve, increasing $\hat{P}^*$ but reducing r^* and g^*. Thus an improvement in the bargaining position of workers, which increases the rate of inflation and reduces the growth rate, could be a cause of stagflation; it can be caused, in a more general model, by other types of supply shocks.

4.4 Kalecki-Steindl models

4.4.1 Money and inflation in Kalecki

Kalecki paid more careful attention to the creation of credit money by banks than did Keynes. He argued that an increase in planned investment increases the demand for credit by firms, and this additional credit is granted by banks; the supply of money is thus endogenous. This is in sharp contrast to the neoclassical view of the exogeneity of money supply, and the Kaleckian position has found acceptance in some of the post-Keynesian literature.[26] It has sometimes been argued that the central distinction between the neoclassical approach and the Kaleckian approach is the fact that in the former money is treated as outside money which can be, and is directly controlled by the monetary authorities, while the latter takes into consideration the existence of credit money. We have seen in section 4.1.2, however, that the neoclassical approach is easily able to incorporate credit money, by introducing (4.5) and assuming given cu and re; this fixes $\hat{M}$ if $\hat{H}$ is fixed. The Kaleckian approach differs from this by making $\hat{M}$ endogenous, despite the possible exogeneity of $\hat{H}$, and this means that cu and re (particularly the latter) must be allowed to adjust. Thus Kalecki in effect argued that the rate of growth of the money multiplier (the term in (4.5) within square brackets) is endogenously determined. He did not provide a detailed theory of the workings of the banking system which would generate such endogeneity, apart from the claim that the supply of credit was demand determined; the behaviour of banks must be examined more carefully to provide a convincing 'microfoundation' for his assumption. A start has been made by Rousseas (1986) who applies Kalecki's markup theory to the banking sector: banks operating with excess reserves (the analogue of excess capacity) set the interest rate as a stable markup (depending on the degree of monopoly in the banking sector) on the discount rate (the analogue of the prime cost), and supply credit according to demand, so that the supply of money becomes endogenous. Even if this analysis is found to be too simplistic, there is no reason to assume the constancy of the money multiplier as in the neoclassical approach.[27]

The endogeneity of $\hat{M}$ implies, as in the neo-Marxian and neo-Keynesian approaches, that there must be a theory of inflation other than the quantity theory. Kalecki did not present a fully developed treatment of inflation, but his works show that he distinguished between conditions of hyperinflation and normal inflation. During conditions of hyperinflation the quantity theory, with inflation depending on the rate of growth of money supply, applies (Kalecki, 1962). But, during normal situations, inflation occurs due to increases in money wages, given the degree of monopoly power.

4.4.2 Cost-push inflation in a Kalecki-Steindl model

This idea can be formally introduced into the Kalecki-Steindl model by using the usual Marxian wage adjustment equation (4.8). The rest of the model is given by (2.1), (2.3), (2.4), (2.13), and

$$g = g(r + \hat{P}, X/K) \tag{4.18}$$

which generalizes (2.14) to allow for capital gains.[28]

The determination of equilibrium is shown in Figure 4.5. From (2.3), (2.13), and (4.18) we get

$$g = g(r + \hat{P}, r(1 + z)/z) \tag{4.19}$$

which implies a positive relation between g and r for a given $\hat{P}$, depicted by curve GG. The goods market equilibrium r for a given $\hat{P}$ is determined where this curve intersects the saving line sr. A higher $\hat{P}$ pushes up the GG curve, resulting in a higher equilibrium level of r, so that we get a positively sloped IS curve. Equation (2.13) implies, for given z and a_0, that (4.10) holds. Substituting this and (2.13) in (4.18) implies

$$\hat{P} = \Omega[\tilde{V} - 1/(1 + z)a_0]. \tag{4.20}$$

Given $\tilde{V}$, we determine the value of $\hat{P}$, which is shown by line PP in the figure, and its intersection of the IS curve determines r^*; the lower diagram then determines g^*. Thus $\hat{P}^*$ is determined completely by the Marxian wage adjustment mechanism, while g^* is determined through the interaction of inflationary forces and the saving-investment parameters.[29]

An increase in $\tilde{V}$ pushes up the PP curve and increases $\hat{P}^*$ and g^*. An increase in z – reflecting a higher degree of monopoly power – shifts the desired rate of accumulation down and hence the IS curve to the right, and the PP curve up. $\hat{P}^*$ necessarily rises, but the effect on r^* is ambiguous, with a higher Ω_1 increasing the possibility of a positive impact. Since the direction of change in g^* in this case is the same as that of r^*, it is possible that a lower markup may not necessarily increase the rate of growth, bringing into question the

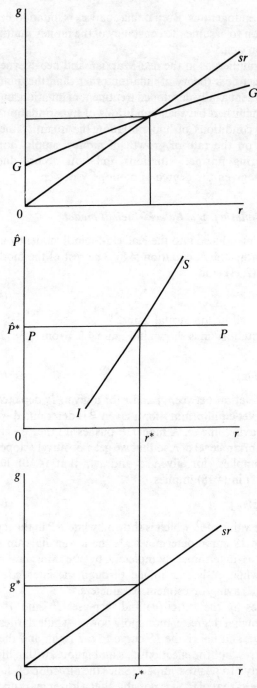

Figure 4.5   A hybrid Kalecki-Steindl–neo-Marxian model

robustness of the proposition established in section 2.2.4. The possibility of g^* increasing with z arises here because investment depends on $\hat{P}$, and will not arise if (2.14) replaces (4.18) and the *IS* curve is vertical.

4.4.3 Conflicting-claims inflation in a Kalecki-Steindl model

A feature of the model just discussed is that the markup rate is constant, given perhaps by the level of industrial concentration, which implies that an increase in the money wage is passed on entirely to consumers in the form of a higher price. Kalecki's (1971a) discussion of the determinants of the degree of monopoly power, however, suggests that it can be affected by the strength of unions, which presumably affects V.

To allow z to be determined by both the level of concentration and the strength of unions, we modify the theory of inflation by assuming that firms cannot completely pass along wage increases. For workers we assume (4.8), rewritten as

$$\hat{W} = \Omega_1 [V_w - V], \tag{4.8''}$$

where V_w is the real wage targetted (desired) by workers, based on the power of unions. For firms, we assume that they have a target (or desired) rate of markup, z_f, which depends on their perception of their degree of monopoly power in product markets. This implies, from (2.13), a targetted real wage, V_f. Whenever the actual real wage is higher than V_f (that is, their desired markup is greater than their actual markup), the firms push up the price according to the adjustment equation

$$\hat{P} = \Omega_2 [V - V_f]. \tag{4.21}$$

The rate of change in the real wage, $\hat{V}$, from (4.14), (4.8''), and (4.21) is

$$\hat{V} = \Omega_1 [V_w - V] - \Omega_2 [V - V_f]. \tag{4.22}$$

The equilibrium value of the real wage is thus seen to be

$$V^* = [\Omega_1/(\Omega_1 + \Omega_2)]V_w + [\Omega_2/(\Omega_1 + \Omega_2)]V_f. \tag{4.23}$$

Its determination is shown in Figure 4.6, where the $\hat{W}$, $\hat{P}$, and $\hat{V}$ (the difference between $\hat{W}$ and $\hat{P}$) lines depict, respectively, equations (4.20), (4.21), and (4.22). The arrows show the movement of the real wages at different levels of V, and attest to the stability of the adjustment process.

The equilibrium values V^* and $\hat{P}^*$ are determined in the figure. We assume that the $\hat{W}$ and $\hat{P}$ lines intersect at $\hat{P} > 0$, to ensure $\hat{P}^* > 0$; this requires $V_w > V_f$, which implies that the shares claimed by the two groups – workers and capitalists – are conflicting. Once V^* is solved, substitution in (2.13) solves for z^*. Substitution of $\hat{P}^*$ and z^* in (2.4) and (4.19) solves for g^* and r^*, as depicted in the lower part of the figure, where GG represents (4.19) after substituting for $\hat{P}^*$ and z^*.

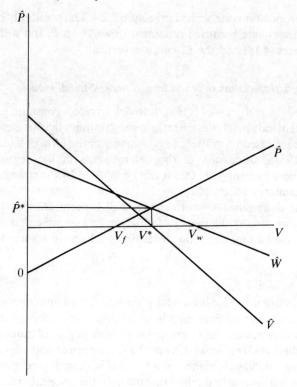

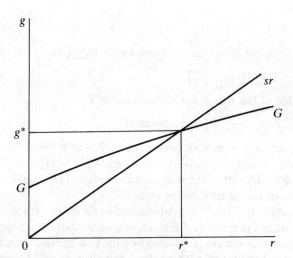

Figure   4.6   A hybrid Kalecki-Steindl–neo-Marxian model with conflicting-claims inflation

Note that V^*, and hence z^*, as (4.23) shows, is determined by the real wages desired by workers and firms, and the coefficients of adjustment in the wage and price change equations. Thus the long-run value of z depends not only on forces relating to industrial concentration as emphasized by Kalecki, but also on class struggle forces emphasized by Marx, which Kalecki (1971a) mentioned in his discussion of the effects of union power on the markup.

We consider, in turn, the effects of greater concentration (lower V_f), and greater union power (higher V_w).[30] A fall in V_f pushes up the $\hat{P}$ curve, increasing $\hat{P}^*$, but reducing V^* and hence raising z^*. These changes leave the *sr* curve unchanged, but the effect on the desired accumulation function is uncertain. The increase in z^* will push it down, while the increase in $\hat{P}^*$ has the opposite effect: the net effect depends on the relative sizes of the two derivatives of the desired accumulation function. If $\hat{P}$ did not enter as an argument, a higher degree of monopoly power would necessarily shift it down, and reduce g^* and r^*, as in section 2.2.4. Thus an increase in the degree of monopoly power will increase $\hat{P}^*$, and either reduce (the absence of $\hat{P}$ as an argument in the g function is a sufficient condition for this) or increase g^*. The case in which a higher $\hat{P}^*$ accompanies a lower g^* can be called the case of stagflation. An increase in V_w shifts the $\hat{W}$ curve up, increases $\hat{P}^*$ and V^*, and reduces z^*. The effect is an upward shift in the desired accumulation function, so that the higher $\hat{P}^*$ is accompanied by a higher g^*.

One interesting implication of this model is that a higher rate of wage indexation increases the rate of inflation, and also increases the rate of growth. To see this, rewrite (4.8″) as

$$\hat{W} = \Omega_1[V_w - V] + \Omega_3 \hat{P}, \tag{4.24}$$

where $\Omega_3 < 1$ is the rate of indexation. In long-run equilibrium, when (4.10) is satisfied, we get

$$\hat{W} = \Omega_1/(1 - \Omega_3)[V_w - V]$$

which takes the same form as (4.8″), with Ω_1 replaced by $\Omega_1/(1 - \Omega_3)$. A rise in Ω_3 thus has the same effect as a rise in Ω_1 in (4.8″), from which our conclusion regarding growth and inflation follows.[31]

The model developed here can be extended to allow for endogenous changes in V_w and V_f.

A natural assumption to make about V_w – following both mainstream approaches which follow the Phillips curve, or Marxist approaches which explicitly introduce bargaining power (Rowthorn, 1980, for example) – is to make changes in V_w depend negatively on the rate of unemployment in the economy. If V_w falls due to a high level of unemployment, as shown, *ceteris paribus*, z^* will increase, resulting in a fall in g^* and also the employment rate (given a constant rate of growth of labour supply). This higher unemployment rate leads to another round of falls in V_w and g^*, so that the economy is caught in a downward spiral. Downward wage flexibility, instead of solving the unemployment problem, aggravates it.[32]

The complete story depends also on how V^f changes over time, and, as discussed in section 3.4.2, there are several different possibilities. If a low rate of capacity utilization reduces it, the economy may be provided with a stabilizing influence, since in the downturn the fall in V_f will help to reduce z^*, raise aggregate demand, the growth rate, and capacity utilization (and conversely in the upturn). Combined with the dynamics in V_w this could result in cycles. If, however, a low rate of capacity utilization increases V^f, the downward spiral will be exacerbated; but, if the demand constraints eventually reduce V_f, this spiral will be checked. Not much more on this than these speculations can be offered at this stage.

4.5 Conclusion

This chapter has introduced money into the alternative models of growth and income distribution, and analysed the causes and role of inflation in them. Our treatment of money has been quite cursory, and the models not much more than illustrations of the ways in which our basic models can be modified to incorporate inflation.

We have argued that there are two different approaches to the treatment of money. The neoclassical approach treats its supply as exogenous, while the neo-Marxian, neo-Keynesian, and Kalecki-Steindl approaches make it endogenous by emphasizing the hoarding and dishoarding of money (in the Marxian approach) and the behaviour of financial institutions (especially in Kalecki's work).

Corresponding to these different approaches to money, there are differences in the explanation of inflation. The neoclassical approach makes the rate of inflation depend on the exogenously fixed rate of increase of money supply. Since the other approaches make this rate endogenous, they invoke other mechanisms such as money-wage dynamics, demand-supply imbalances, and conflicting claims. We have incorporated these mechanisms into some of our basic models, and sometimes introduced conflicting features (or closing rules) from different basic models to develop hybrids in which inflation resolves these conflicts. We have shown, however, that the different models have strikingly different implications on the relationship between growth, distribution, and inflation, so that careful attention must be given to how, and in what basic models, these features are introduced.

CHAPTER 5

Technological change

5.0 Introduction

We have assumed so far that the input–output coefficients a_0 and a_1 (or when factor substitution is allowed, the relationship between them) are given. In the neoclassical approach this is tantamount to assuming a given technology. In the other approaches – especially the neo-Marxian one – the coefficients can depend on factors other than technology, such as social relations of production. But there too, given coefficients imply a given technology, unless changes in social relations and technology have the accidental and implausible effect of offsetting each other.

The assumption of given technology is clearly inadequate for a proper analysis of growth and distribution. First, a large number of economic historians and theorists emphasize the *empirical* importance of technological change in the growth process, so that the study of growth seems to be at best incomplete without it. Secondly, some of the approaches *theoretically* view technological change to be integrally related to the process of growth and determination of income distribution, so that the assumption of given technology in models representing these approaches thus does them serious injustice. The primary example is the Marxian one: no exposition of neo-Marxian models is complete with given technology. Among neo-Keynesians, the role of technological change in the growth process has been emphasized by Kaldor. Kalecki, especially in his later writings, also stressed the role of technological change in generating investment demand. Even neoclassical growth theory has incorporated technological change to make the models more realistic.

The four sections of this chapter will incorporate technological change into our four basic models, and examine its consequences. We will examine simple, illustrative, ways in which technological change can be, and has been, introduced into them, but make no attempt to provide a thorough examination of the role of technological progress in economic growth. We will offer no systematic treatment of the causes or nature of technological change, focusing rather on its consequences for growth and income distribution. Further, several important approaches to growth and technological change, including the Schumpeterian and evolutionary approaches, will not be discussed. The

87

Schumpeterian approach has proved hard to model mathematically, and does not fit into any of the approaches that have been examined in chapter 3. The evolutionary models are either informal or, in the hands of Nelson and Winter (1982), show non-equilibrium and non-steady state behaviour, so that they are difficult to fit into our framework; they are also geared more towards explaining technological change rather than exploring its growth and distributional consequences.

5.1 Technological change in neoclassical models

The neoclassical models explored in sections 2.2.1 and 3.1 imply that in long-run equilibrium the labour-output ratio, a_0, is constant, implying that per-capita output is constant. This is somewhat of an embarrassment if we use long-run equilibrium as a device with which to study actual economies, since actual economies usually experience growth in per-capita income. Moreover, while in the models growth is the result only of labour supply growth and capital accumulation, empirical work on the sources of growth has found the contribution of technological change to be of great importance.[1] Neoclassical models, however, can be easily modified to incorporate technological progress to solve the second problem; this, moreover, solves for the first problem, since with appropriate types of technological progress it is possible to examine long-run equilibria in which per capita income increases.

5.1.1 Alternative approaches to technological change

The literature on technological change in neoclassical growth models is immense. Several types of models have been studied, and can be distinguished according to the following criteria:

1   There are some models which assume technological change to be exogenous, occurring at a prespecified rate, falling like manna from heaven, and those which take it to be endogenous, depending on economic factors examined within the model. While to have a complete analysis of the interaction of growth and technological change the second approach needs to be adopted, there is a strong case – on account of its simplicity – for the study of exogenous technological change, at least as a starting point. If endogenous technological progress is assumed, there is the further issue of what it depends on.

2   Technological change may be modelled as disembodied (changing the relationship between inputs and the output without changing the qualitative nature of the input), or embodied (transforming the nature of the inputs so that, for instance, capital of different vintages have to be distinguished). Both types are worth studying, since they reflect different facets of technological progress: the former, organizational improvements, and the latter, better machines and better-trained workers.

3 There is the question of the *bias* in technological change, that is, whether it is primarily labour saving, capital saving, or neutral. Several different notions of neutrality have been used. Restricting ourselves to constant returns to scale technology, we may distinguish between three notions of neutrality: if, before and after technological change, cost-minimizing firms choose the same capital-labour at a given wage-rental ratio we have Hicks-neutrality, if they choose the same capital-output ratio at a given rental rate we have Harrod-neutrality, and if they choose the same labour-output ratio at a given wage we have Solow-neutrality. Corresponding to each notion of neutrality are notions of biased technological change: for example, technological change is biased in the direction of being capital (labour) using in Harrod's sense if at the same rental rate the capital-output ratio increases (decreases).

Apart from these distinctions, the neoclassical models differ according to what they assume about the structure of the model; for example, whether they allow for factor substitution (and in what way).

Particular models may make independent choices regarding each one of these different criteria, although some may try to derive answers about some of them based on choices made regarding others.

5.1.2 Exogenous technological change

A particularly simple model introducing technological change into a neoclassical model and yielding growing per-capita output in long-run equilibrium is developed by assuming exogenous, disembodied, and Harrod-neutral technological progress at a given rate into our neoclassical model of section 2.2.1. This amounts to assuming that a_0 falls over time at a constant rate, h, so that

$$\hat{a}_0 = h, \tag{5.1}$$

while a_1, the capital-output ratio, does not change.

In this case, we can still use equations (2.1), and (2.3) through (2.5) but (2.6) no longer holds. To see this, recall that (2.6) shows growth equilibrium with full employment of labour, which requires that the demand and supply of labour grows at the same rate. Now the supply of labour is assumed to grow, as before, at rate n. But the demand for labour, instead of growing at rate g, grows at

$$\hat{L} = \hat{a}_0 + g$$

since $L/X = a_0$, and since the capital-output ratio is constant in (2.5). Equating the rate of growth of labour demand to the rate of growth of labour supply, and using (5.1) we obtain the full employment growth condition,

$$g = n + \hat{a}_0. \tag{5.2}$$

Given n and h, (5.1) and (5.2) solve for g^*. Given g^*, (2.4) solves for the equilibrium rate of profit. With K/X determined from equation (2.5), $g^*(K/X)^*$

and $r^*(K/X)^*$ are seen to be constant, so that (2.1) and (2.3) determine the equilibrium values of the products Ca_0 and Va_0. Since these products are constant in equilibrium, it follows from (5.1) that

$$\hat{C}^* = h \qquad (5.3)$$

and

$$\hat{V}^* = h. \qquad (5.4)$$

Unlike the earlier neoclassical models, consumption per worker and the real wage grow at a constant rate, at the rate of decline in the labour-output ratio. Note that, although the real wage rises along the growth path, since $Va_0/r(K/X)$ is constant, the labour share in total income does not change for given values of the parameters. A faster rate of technological change, that is, a higher h, implies, by (5.2), a higher g^*, hence by (2.4) a higher r^*, by (2.3) a lower labour share in total income, but by (5.3) and (5.4) a higher $\hat{C}^*$ and $\hat{V}^*$.

It is thus a simple matter to incorporate technological change into a neoclassical model and examine the properties of growth equilibrium. Moreover, in equilibrium per-capita income (strictly, per-worker income) grows at a constant rate $g - n = h$ rather than being constant. But all this is true for a particular neoclassical model and for a special type of technological change.

Similar results can be obtained from other types of neoclassical models. Indeed, the first neoclassical models with technological change assumed substitution in production rather than fixed coefficients. If we assume that the production function is given by (3.5) with the usual neoclassical properties, and we assume that technological change is Harrod-neutral, the equilibrium rate of growth of the economy is given by $n + h$, which is the rate of growth of the supply of labour in efficiency units.

The nature of technological change, that is Harrod-neutral and exogenous, still remains specific. There have been attempts to explain Harrod-neutrality by using approaches which allow firms to choose the nature of technological progress. Following Kennedy (1964), Drandakis and Phelps (1966), and Samuelson (1966), assume that the representative firm has a production function given by (3.5) with usual neoclassical properties, and that there is a given function, called the innovation possibility function, T,

$$-\hat{a}_0 = T(-\hat{a}_1), \qquad (5.5)$$

where T', $T'' < 0$ and there are upper limits to $-\hat{a}_0$ and $-\hat{a}_1$, which represents the trade-off faced by the firm in achieving rates of growth of labour and capital productivity. Assume that firms choose the point on the innovation possibility function which maximizes the rate of growth of output at given factor prices. Then it can be shown that at steady state equilibrium (at which the growth rate and the rate of profit take constant values), technological progress will be such that $\hat{a}_1 = 0$ and $-\hat{a}_0 = T(0)$, which implies Harrod-neutrality.[2] It thus appears that the nature of technological progress, that is, its Harrod-neutrality, is no longer assumed in this model, but is explained within it. It should be noted, however, that technological progress is still exogenous, since the innovation possibility function is given to the firm irrespective of what it does, allowing it only to choose the bias of technological change.

5.1.3 Endogenous technological change

There have been several attempts to introduce endogenous technological change in neoclassical models. One approach assumes that firms (or the economy) have to incur expenses for raising productivity, either for research and development, or for training and education.[3] This approach essentially makes technological change similar to capital accumulation: additions to both the stock of technology and capital have to come from savings. Another approach assumes that producers learn to produce better by gaining production experience, or what is called, following Arrow (1962), 'learning by doing'. In this approach saving does not have to be devoted to 'investment' for technological change: it occurs simply as a result of experience. The two types of endogenous theories of technological change capture two different mechanisms through which technological change actually occurs in economies,[4] both undoubtedly important. Neoclassical models incorporating these mechanisms have been used to examine whether under perfect competition (including perfect foresight) an economy can grow optimally, in the sense of maximizing a social welfare function. The general answer has been in the negative, due to the existence of externalities related to the mechanisms which optimizing agents do not take into account, and 'optimal' interventions for the removal of these distortions have been examined.

The first type of model need not detain us since it is a close variant of models which grow due to capital accumulation from saving, with few additional features.[5] Instead, we consider briefly a learning by doing extension of the basic neoclassical model, to a variant of which we will return when we discuss uneven development in chapter 10.

Arrow's (1962) model assumes that technological change due to learning is purely labour augmenting, and is embodied in new machines. Arrow measures learning by cumulative investment which, in the absence of depreciation, is equal to the stock of capital. To simplify, we assume that technological change is disembodied, but otherwise follow his characterization. Assume that the learning function takes the specific form

$$1/a_0 = AK^m \tag{5.6}$$

with $m < 1$ due to diminishing returns to learning, and with A and m constants.[6] This implies, upon logarithmic differentiation,

$$-\hat{a}_0 = mg. \tag{5.7}$$

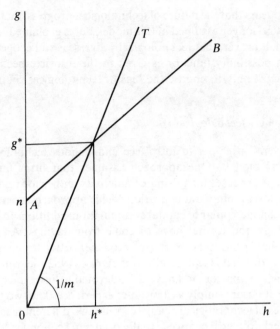

Figure   5.1   A neoclassical model with learning by doing

Consider now the model given by (2.1), (2.3), (2.4), (2.5), (5.1), and (5.2). Instead of $-\hat{a}_0$ (or h) being fixed in (5.1) and (5.2), it is now a variable, and (5.7) closes the model. The determination of equilibrium for this model is shown in Figure 5.1. AB shows an upward rising relation between g and h given by equation (5.2); its intercept is n and its slope is 1. OT represents equation (5.7) (using (5.1)), and has a slope $1/m$. Their point of intersection determines g^* and h^*. With g^* solved, the equilibrium values of all variables can be solved in the same way as was done for the model with exogenous technological change. Note the importance of the assumption of $m < 1$ for the existence of equilibrium. With $m > 1$, or increasing returns to learning, the growth process becomes explosive. Note also that a rise in m increases the rate of technological progress and growth rate; a rise in n has the same effects. This implies that these changes increase the rate of growth of consumption per worker and real wages, but reduce the labour share.

5.2 Technological change in neo-Marxian models

5.2.1 Technological change in Marx's scheme

While in neoclassical growth theory technological change can be thought of as a modification of growth theory with given technology, in Marx's system technological change is the fundamental force behind change in history.

Capitalists may be expected to innovate to increase their profits. Actually, to survive the competitive struggle against others, individual capitalists are forced to accumulate and innovate. Additionally, the saturation of product markets (at existing prices) may induce capitalists to introduce new products, which requires innovation. Finally, increases in wage costs due to the diminution of the reserve army of the unemployed can reduce profits, inducing firms to cut costs by innovating.

5.2.2 The rising organic composition of capital

Marx also argued that the direction of technological change will be labour saving and capital using. The Marxian law of the rising organic composition of capital, which incorporates this tendency, has generally been assumed and not very carefully explained. Several attempts at explanation, however, exist, including some by Marx.

First, Dobb (1940) and Sweezy (1942) argue that labour shortages resulting in an upward pressure on wages induce firms to adopt labour-saving and capital-using innovations. The argument can be given a neoclassical interpretation, and Dobb in fact explicitly refers to Hicks's (1932) argument, which involves the use of isocost and isoquant lines. Elster (1983) echoes Samuelson's (1965) argument that such a view is incorrect; firms will choose techniques given technology to minimize costs, implying that labour and capital will be just as cheap or as expensive in production, so that there will be no motivation for biasing technological progress in any direction, only in reducing costs. In addition to the issues raised in the earlier discussion (see Samuelson, 1965), this critique overlooks the fact that the distinction between choice of technology and technique is not quite clear-cut; without perfect substitution in the short run it is quite legitimate for firms expecting an increase in wage costs to adopt labour-saving and capital-using innovations. This is particularly so if with a given technology production takes place with fixed coefficients, which is in fact Marx's assumption.[7]

Secondly, and more faithful to Marx himself, Marglin (1976), Wright (1977), and Lazonick (1979) argue that increasing mechanization serves as a weapon against labour, both by reducing the overall dependence of the capitalist on labour, and also by reducing the role of labour in controlling the speed of production.

Thirdly, Heertje (1977) emphasizes that capital intensity may follow from the process of concentration and centralization which results from the competitive process itself. As firms drive out other firms, they may not just become more efficient and reduce costs in general, but may make the technology of production highly capital intensive to keep out potential entrants.

Finally, Harris (1986) argues that as a result of the process of capitalist development, different sectors of the economy may develop differently. The result of this uneven development may be to increase the weight of those sectors with a higher capital intensity. While this argument is appealing, and can be supported by casual empiricism, it remains to be developed more fully.

These arguments are plausible and suggest that technological change is labour saving. But it is more difficult to accept that it is also capital using. Blaug (1960) argues that the empirical evidence does not support that, and that even Marx, unlike his contemporaries, recognized the capital-saving nature of many innovations. It may be argued that Marx thought that greater capital requirements were the price that capitalists had to pay for reducing labour use, and the spread of mechanization served to corroborate this impression. While capital needs may have increased, this does not imply higher capital requirements per unit of output.

5.2.3 Law of the falling rate of profit

One of the most important implications of the rising organic composition of capital in Marxian analysis is the law of the falling rate of profit which, according to Marx, implies the eventual collapse of capitalism. A great deal of theoretical and empirical controversy has surrounded this issue, and present appraisals are critical of the law. Some have argued that if the organic composition rises the profit rate need not necessarily fall; others have gone further to argue that the rate of profit cannot fall due to technological change. We examine this issue briefly by introducing technological change parametrically into our basic neo-Marxian model of chapter 2.

Consider the price equation for the neo-Marxian model, which can be written, using (2.3) and (2.5), as

$$1 = Va_0 + ra_1 \tag{5.8}$$

which implies that

$$r = (1 - Va_0)/a_1 \tag{5.9}$$

which can be written in two alternative forms,

$$r = [(1/a_0) - V]/(a_1/a_0), \tag{5.9'}$$

$$r = [(1/a_0 V) - 1]/(a_1/a_0 V). \tag{5.9''}$$

The organic composition of capital is measured alternatively by $a_1/a_0 V$, the value composition of capital, and a_1/a_0, the physical organic composition of capital or the capital-output ratio. The two expressions (5.9') and (5.9'') differ in having alternative notions of the organic composition in the denominator. Other things constant, they show that an increase in the organic composition of capital reduces the rate of profit.[8]

However, the expressions also show that other things may not be constant. Suppose that a fall in a_0 and a rise in a_1 make the organic composition of

capital (by both definitions) rise, making the denominators in (5.9') and (5.9")
rise. But the numerators in both will also rise due to the fall in a_0, and it is not
possible to determine the direction of change in r.[9]

It now becomes necessary to distinguish between the two neo-Marxian
models that were distinguished in section 3.2, one with an exogenously fixed
real wage $\tilde{V}$, and the other with a given rate of exploitation, ϵ. In our earlier
analysis with given technology it did not matter whether $\tilde{V}$ or ϵ was held
constant, but now, with a_0 allowed to change, the two models have different
implications.

For the model with the given ϵ, (3.10) implies that Va_0 is constant, so that
V must rise at the same rate at which a_0 falls due to technological change. The
value composition of capital still increases since a_1 is assumed to rise, so that
the organic composition of capital rises by both definitions. But with a_0V
constant, (5.9") shows that r must fall. Thus the Marxian position on the falling
rate of profit is valid. The problem is to justify the assumption of a given ϵ when
technology changes. By what mechanism does V increase when a_0 falls? Does
the fall in a_0 cause V to rise or is the direction of causation the other way
around? It should be noted that for r to fall the condition that the organic
composition of capital increases and that ϵ does not increase is a sufficient one,
and r may fall even for moderate increases in ϵ. If r falls, by (2.4) so must g. The
effect on $\hat{X}$ depends on whether g falls more or less than $-\hat{a}_1$ (since $\hat{X}=g-\hat{a}_1$),
and the effect on labour absorption depends on the rate of growth of output
plus $\hat{a}_0$ (since $\hat{L}=\hat{X}+\hat{a}_0$).

For the model with a given V, an increase in the organic composition of
capital is not sufficient to reduce r, as (5.9') and (5.9") show. The possibilities
can be illustrated in Figure 5.2. From (5.8), for given V and r, we draw
a relationship between a_0 and a_1 shown as TT, called the isoprofit (rate) curve
and represented by

$$a_1 = (1/r) - (V/r)a_0.$$

Suppose the economy is initially at point A. Technological change of the type
described by Marx implies a movement in the North-West direction. If the
change is along the arrow which keeps the economy on line T, with a given V it
implies that the economy has the same r. If the economy moves along the
arrow which takes it below (above) TT, it implies that a_1 is rising less (more),
for a given fall in a_0, than in the previous case, and, if this happens, r must rise
(fall) for a given V. Thus, only for certain types of technological change, in
which a high price (in terms of a rise in a_1) must be paid for a given fall in a_0,
will r fall with a rising organic composition of capital.

Although this model takes V as given parametrically, there may be forces
which increase V over time (due to a diminution of the reserve army caused by
a growth in the demand for labour), and, as mentioned above, that may be the
motivation for technological change. A rise in V rotates the isoprofit curve
downwards, anchored on its vertical intercept, making it steeper. The new

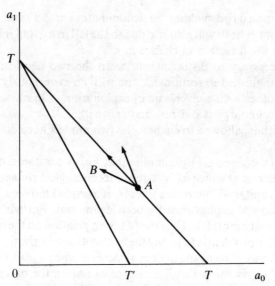

Figure 5.2   Technological change in the neo-Marxian model

curve is shown in Figure 5.2 by TT'. If the economy stays at A, it will have a lower r, since the economy is above its isoprofit curve. It is also obvious from the figure that a technological change which could previously have resulted in a higher r (taken the economy below TT to say B) can now imply a lower r (above TT') due to the wage increase. Thus an increase in V increases the chances of a fall in r.

A full analysis of whether or not the rate of profit will fall over time requires not only a confirmation of whether the organic composition rises over time, but also information on the nature of technological change and the behaviour of the real wage. Harris (1983, 1986) and Glombowski (1986) provide a flavour of what is involved, and the general conclusion that follows is that the Marxian law regarding the falling rate of profit cannot be proven, even assuming an increasing organic composition of capital. It depends on such characteristics of the economy as the responsiveness of technological change to the other variables of the model, the responsiveness of real wages to the size of the reserve army, and the capacity of capitalists to tap new sources of labour supply.[10]

Some writers have gone beyond the argument that the rate of profit need not fall, and have insisted that it cannot.[11] The basic proposition, due to Okishio (1961, 1977), is that if capitalists introduce technical innovations when and only when it is cost reducing at existing prices and the profit rate, if perfect competition prevails, and if the real wage remains constant, technological change must raise the equilibrium rate of profit. In terms of Figure 5.2, it is being argued that capitalists will accept only those innovations (which

Roemer calls 'viable' ones) which are on arrows taking them from A to positions below the TT curve. To see this, assume, following Okishio, that capitalists will mechanize only if it reduces costs at the initial prices. This implies that a change (da_1/da_0) will be accepted if and only if $Vda_0 + rda_1 < 0$, or $-da_1/da_0 < V/r$, proving that only innovations shown by arrows flatter than the TT curve (with slope V/r) will be accepted.[12] The proposition has been criticized severely by a number of writers, including Shaikh (1978, 1980) and Laibman (1982).

Instead of entering into the controversy in detail, let us note that, while the theorem is true, it does not imply that the rate of profit in a capitalist economy cannot fall, since all its assumptions need not be valid. First, it assumes that firms are perfectly competitive in the neoclassical sense of being price takers. If this is not the case, the firms will make decisions based on some profit expectation and, as Shaikh (1980) argues, there is no reason why the equilibrium rate of profit must rise when the technological change occurs. Secondly, it assumes maximizing behaviour in an environment where the technical effects of the changes are known; if the outcome of innovational efforts is uncertain, and this may be because such changes are in many cases not entirely intentional (learning by doing, for example), as Roemer (1979) admits, the final outcome cannot be known by firms. Finally, as mentioned above, technological change, by raising the labour productivity, can raise the real wage; in this case the real wage change is not exogenous to the technological change and is induced by it, an effect which may be ignored by the innovating firms (since it is a macro effect). Laibman (1982) has argued that rather than assuming that V is given, as the Okishio theorem does, it is more 'neutral' to take ϵ (which depicts the state of the class struggle) to be given, so that V rises when technology changes. In this case, even if capitalists undertake only viable changes, r may fall.

5.2.4 The role of technological change

We conclude with a general comment on the role technological change plays in the neo-Marxian model. Since labour is not fully employed, the effect of technological progress is not to raise growth by increasing the effective supply of labour as in the neoclassical model. To make a comparison, consider the simple case in which the organic composition of capital increases over time only due to a fall in a_0 with a_1 constant (the case of Harrod neutral technological change), and a given rate of exploitation ϵ (with V rising with labour productivity).[13] In this case equation (5.9) shows that r does not change, so that we have an equilibrium like those we considered in the neoclassical model. But, in this model, with r stationary, so is g, so that given a_1, the rate of growth of output does not change. The rate of growth of employment is then given by

$$\hat{L} = g - h, \tag{5.10}$$

where h is the given rate of fall in a_0, which shows that a higher h, since it leaves g unchanged, will reduce $\hat{L}$. This does not imply that technological change cannot affect the rate of growth in the model. We have already mentioned how it may do so – in an adverse manner – if it reduces the r due to its effect on a_1. Another possibility is that by reducing the rate of labour absorption, and (given the rate of growth of labour supply) thereby increasing the unemployment rate, it can increase ϵ, raise r, and increase the rate of growth.

5.3 Technological change in neo-Keynesian models

5.3.1 The role of technological change

If we assume that the economy experiences Harrod neutral technological change at rate h, as given by (5.1), but maintain all the assumptions of the neo-Keynesian model made in section 2.2.3, we find that g is determined by the desired accumulation function and the saving ratio. Thus g and r are not affected by the rate of technological progress, and, with a constant capital-output ratio, nor is the rate of growth of output. The real wage, given in section 2.2.3 by

$$V = (1 - r^* a_1)/a_0$$

increases at rate h since a_0 declines at that rate. The rate of absorption of labour is given once again by equation (5.12). It is easy to see that with Harrod-neutral technological change $a_0 W/P$ is constant, so that the labour share does not change with technological change: the greater unemployment is compensated by the higher real wage.

If technological change is not Harrod-neutral, but changes a_1 as well, g is still independent of the rate of technological change. But we now have

$$\hat{X} = g - \hat{a}_1. \tag{5.11}$$

If technological change reduces the capital coefficient, so that $\hat{a}_1 < 0$, $\hat{X}$ will increase with a higher $\hat{a}_1$, and, with r at its equilibrium value, the labour share will increase over time and the capital share will fall. Labour absorption is given by

$$\hat{L} = g - \hat{a}_1 + h. \tag{5.12}$$

With g unaffected by changes in the rate of technological change, the effect on $\hat{L}$ depends on the strength of $\hat{a}_1$ and $\hat{a}_0$. There may be a further effect on the growth of output – through changes in g as well – if the rate of technological change enters as a parameter in the desired accumulation function, a possibility that will be explored in section 5.4.

5.3.2 Kaldor on the role of technological change

In contrast to our conclusions, Kaldor (1955–6, 1957, 1959, 1961), the leading analyst of technological change in the neo-Keynesian tradition, provides a very different analysis. In his analysis the fundamental determinants of the long-run rate of growth are the rate of growth of labour supply and the technological dynamism of the economy. Greater technological dynamism always increases the long-run rate of growth, as in (5.2) for the neoclassical model. That Kaldor obtains this neoclassical result, despite his neo-Keynesian background (he uses the desired accumulation function), may appear paradoxical.

In examining Kaldor's analysis, we begin with his analytical innovation for representing technological dynamism, the technical progress function. Kaldor (1957, 1961) makes a series of criticisms of the neoclassical production function which attempts to portray the efficient relation between levels of output and inputs for a given state of technology. Most important for our purposes,[14] he argues that technical progress and capital accumulation are necessarily interdependent and therefore cannot be separated from each other theoretically or empirically. Rejecting the production function, he introduces the technical progress function

$$X / L = T(K / L) \tag{5.13}$$

with $T(0) > 0$, $T' > 0$, and $T'' < 0$, which shows that the growth rate of output per worker is an increasing function of the growth rate of capital per worker, subject to 'diminishing returns', and output per worker can increase without a growth in capital per worker. This function, illustrated in Figure 5.3, is supposed to capture the relationship between technical change and capital deepening. Kaldor (1957) argues that most 'technical innovations which are capable of raising the productivity of labour require the use of more capital per man – more elaborate equipment and/or mechanical power'. Moreover, there can be learning effects and dynamic economies of scale associated with the production and the use of capital goods. Kaldor (1966) examined such effects in his empirical studies of Verdoorn's Law. Note also the similarity between Kaldor's function and Arrow's learning function discussed earlier: the latter makes the rate of growth of labour productivity depend on the rate of growth of capital, while the former makes it depend on the rate of growth of capital per worker.

Kaldor sometimes uses a linear form given by

$$X / L = T_0 + T_1 (K / L) \tag{5.12'}$$

with $T_i > 0$. Rewriting it as

$$\hat{X} = T_0 + T_1 \hat{K} + (1 - T_1)\hat{L}$$

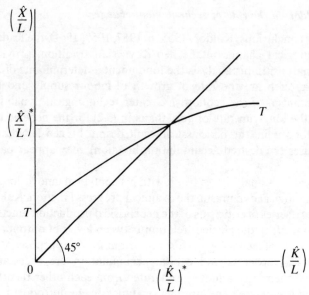

Figure 5.3   Kaldor's model of technological change

and integrating, we get

$$X = C_0 K^{T_1} L^{1-T_1} e^{T_0 t},$$

where t is time, which is easily recognized as a Cobb-Douglas production function with exogenous technical progress at rate T_0. On this basis Black (1962) argues that the technical progress function is no different from a special neoclassical production function. However, two comments on this are in order. First, for Kaldor the function is an empirical relation between two observables and in some estimated non-linear form it may not be integrable, which confirms Kaldor's point that it may not be possible to write down a production function for a given technology; the fact that its linear form yields a Cobb-Douglas production function does not detract from the force of this argument. Secondly, if X/L and K/L are taken to be short-run givens, the function implies fixed coefficients in the short run, but allows for substitution possibilities only with technical change, which makes it different from the neoclassical production function. Note that the use of fixed coefficients production function does not necessarily imply efficiency, or the technological determination of the input–output coefficients.

In his analysis of the determinants of long-run growth, Kaldor argues that the economy is at full employment in the long run. Assuming that at long-run equilibrium the economy must have a constant capital-output ratio (which implies that technical change does not involve changes in the capital-output ratio), we must have $\hat{X} = \hat{K}$, which implies that $\hat{X} - \hat{L} = \hat{K} - \hat{L}$, or

$$X/L = K/L, \tag{5.14}$$

so that the economy must come to long-run equilibrium at the intersection of the 45° line and the TT curve in Figure 5.3, satisfying both (5.14) and (5.13). The rate of growth of labour productivity is thus determined by the technical progress function. In the case of the linear function (5.13') we get

$$X/L = T_0/(1 - T_1).$$
(5.15)

Full employment growth at an exogenously fixed rate n implies $\hat{L} = n$, so we get

$$\hat{X} = g = T_0/(1 - T_1) + n$$
(5.16)

from which Kaldor's conclusion follows: the growth rate of the economy depends only on the rate of growth of labour supply and the technological parameters.

Kaldor sought to defend his assumption of long-run full employment growth on both empirical and theoretical grounds. Empirically, he believed – even in his later writing (see Kaldor, 1986) – in the stylized fact that advanced economies on average experience full employment growth, although they do sometimes experience unemployment. This belief is shared by many, but not by most non-neoclassical economists. Theoretically, he tried to show that the economy will converge to full employment in the long run, though not with neoclassical wage flexibility arguments.

His analysis has two elements. First, he argued that, if the economy is at full employment, changes in saving due to shifts in income distribution caused by price changes with given money wages will maintain full employment. Such changes in saving occur because of differences in the saving propensity out of wage and non-wage income. Secondly, to explain why the economy is at full employment to begin with, Kaldor assumes a specific investment function which makes investment depend on the level of income, and that changes in income cause a greater change in investment than in saving. This implies that excess demand induced increases in output and income will increase excess demand, driving the economy to full employment. He argues that during conditions of growth these assumptions are satisfied, so that in growth equilibrium the economy will always be at full employment.

The validity of Kaldor's argument depends on the nature of his investment function,[15] and several remarks about this are in order. First, Kaldor argues that under conditions of growth changes in investment are induced by changes in output, but this is necessary, but not sufficient, for ensuring that the investment response to output exceeds the saving response; thus, even with his investment function, unemployment equilibrium is possible. Secondly, if his mechanism is accepted, the following problem arises. If output changes are allowed in the short run, full employment will prevail in that run, something Kaldor presumably would not accept. If output changes are not allowed in the short run, then price and the price cost margins will have to change in that run, contrary to what Kaldor seems to believe. Thirdly, the assumption that investment is more responsive than saving to changes in income is contrary to the basic tenet of Keynesian macroeconomics of virtually all shades, which argues that at least in the short run, output is demand determined.[16] Fourthly, even if we accept his mechanism, there is no logical reason for the economy to be pushed to full employment; it could be pushed to the full utilization of (capital) capacity, which may not be enough for the full employment of the labour force.

These comments suggest that it is unrealistic, un-Keynesian, un-Kaldorian, and even logically inadequate, to use Kaldor's mechanism to explain why the economy will have full employment in the long run. Indeed, in his growth models the investment functions actually do not exhibit the characteristic of being more responsive than the savings functions. His models thus *assume* full employment in the long run, rather than *deducing* it from assumptions about investment and saving functions. This seems to be an undesirable way to proceed, especially in view of the fact that the question of full employment is an object of such great dispute between economic theorists of different persuasions.

5.3.3 The role of technological change in a Kaldorian model

It seems preferable to consider a model which does not assume full employment, but allows for the possibility of unemployment. The model we consider is the same as the one developed in section 2.3.4, but append to it the technical progress function in its linear form, given by equation (5.9'). To recapitulate briefly, we consider a one sector economy with a representative firm which uses a fixed coefficient production function. Wage income goes to workers who consume their entire income while profit income goes to capitalists who save a fraction s. Firms make investment plans according to (2.31), with the investment rate depending positively and linearly on both the rate of profit and the output-capital ratio. The money wage is given. The firm possesses market power; the price is either set as a markup over prime costs (if excess capacity exists) or at a level which equates demand and supply (if output cannot be increased due to capacity or labour constraints). Population grows at rate n, and technological progress occurs according to (5.9'). This model is similar in spirit to Kaldor's although it employs some different assumptions, especially a simpler investment function which makes the analysis more transparent.[17]

Assuming, to start with, that the input–output coefficients are given, we have already seen (in section 2.3.4; see also section 2.3.3) that the economy can, in long-run equilibrium, end up in one of three states. First, it can be labour constrained and grow at rate n, with an excess capacity of capital. Secondly, it can be capital constrained, and grow at a rate determined by the parameters of the saving and investment functions; it will fully utilize its capital, but have unemployment of labour. Finally, it can be demand constrained, and grow at a rate determined by the parameters of the saving and investment function and the given markup of price over prime costs for the economy, with both excess capacity and unemployment of labour. Which of the cases the economy will end up in depends on the parameters of the model, and one cannot necessarily conclude that the economy will end up with full employment growth. Here we consider the implications of technological change according to (5.13') in the first two cases, postponing discussion of the demand constrained case to the next section.

In the labour-constrained case, the implications are the same as in Kaldor's analysis, as we would expect. Since in this case we have $X = N/a_0$,

$$\hat{X} = n - \hat{a}_0. \tag{5.17}$$

Using the definition of a_0 we can write (5.13') as

$$-\hat{a}_0 = [T_0/(1 - T_1)] + [T_1/(1 - T_1)] [\hat{K} - \hat{X}]$$

which, in equilibrium growth, with $\hat{K} = \hat{X}$, and using (5.17), implies (5.16).[18]

In this case, in long-run equilibrium (for a given rate of technological change) the rate of profit is fixed, so that the real wage increases over time. A faster rate of technological change increases r^* (since g^* increases, and $g = sr$), raises the rate of growth of the real wage in long-run equilibrium (since $(a_0 V)^*$ must be constant), and also raises $(X/K)^*$ (from (2.4) and (2.31)). But since from (2.34)

$$r(K/X) = [\alpha/(s - \beta)] (K/X) + \tau/(s - \beta)$$

in long-run equilibrium, a higher $(X/K)^*$ implies a lower $r^*(K/X)^*$, implying a higher $(W/P)a_0$ in long-run equilibrium (from equation (2.3)), so that the labour share in income goes up.

In the capital-constrained case, using the definitions of a_0 and a_1, (5.13') implies

$$-\hat{a}_0 = [T_0/(1 - T_1)] - [T_1/(1 - T_1)] (-\hat{a}_1). \tag{5.18}$$

Given T_0 and T_1 we obtain a linear, inverse relation between $-\hat{a}_0$ and $-\hat{a}_1$, similar to the technological function, (5.5), of neo-classical growth theory. While (5.18) cannot tell us where on this technological frontier the economy will be, Kaldor's (1957) stylized facts suggest that a_1 is constant over the long run, so that $\hat{a}_1 = 0$, implying that technological progress is Harrod-neutral. Substituting $\hat{a}_1 = 0$ into (5.18) we get

$$-\hat{a}_0 = T_0/(1 - T_1). \tag{5.19}$$

Since, as (2.34) shows, the equilibrium value of g is independent of the technological parameters (assuming that the investment parameters are independent of them), with constant a_1, the rate of growth of output is also independent of them. Since the rate of growth of employment, $\hat{L} = \hat{X} + \hat{a}_0$, is given by

$$\hat{L} = g - T_0/(1 - T_1) \tag{5.20}$$

greater technical dynamism merely reduces the rate of labour absorption.[19] Our earlier neo-Keynesian result is vindicated even with Kaldor's technical progress function and we may conclude that Kaldor obtained his neoclassical result by assuming full employment growth, and not because of the way he treated technological change.

5.3.4 Implications of Eltis's technical progress function

Kaldor's technical progress function has been criticized by Eltis (1973) because it implies that the rate of growth of the economy depends only on the rate of growth of labour supply and the parameters of the technical progress function, and is independent of investment and saving parameters. Eltis suggests that better results can be obtained if it is assumed that technical progress depends upon the share of investment in output rather than the rate of growth of capital per worker, so that

$$X/L = T_0 + T_1(I/X). \tag{5.21}$$

Eltis justifies this formulation with the argument that it is more profitable to invent and develop superior methods of production when the rate of investment is high, so that the expected profits from innovational activity are high.

Without entering into the validity of this argument, let us examine the implications of (5.21) for our models. In our notation, it can be written as

$$-\hat{a}_0 = T_0 + T_1(gK/X) \tag{5.22}$$

which shows that any parameter which affects gK/X in our models affects the rate of technological change. In the labour-constrained case of our Kaldorian model,

$$g = n - \hat{a}_0 = s[\alpha + \tau(X/K)]/(s - \beta) \tag{5.23}$$

which shows that in long-run equilibrium

$$(gK/X)^* = [\tau s/(s-\beta)] \{[s\alpha/((n-\hat{a}_0)(s-\beta) - s\alpha)] + 1\}. \tag{5.24}$$

In Figure 5.4 curves TT and AB represent (5.22) and (5.24), respectively, and equilibrium $-\hat{a}_0$ and I/X are determined at E. If α, an investment parameter, increases, (5.24) shows that curve AB shifts up (with the stability assumption $s > \beta$) to say $A'B'$ and equilibrium shifts to E', implying a higher $-\hat{a}_0$, and by (5.23), a higher equilibrium g.[20]

However, it should be noted that Eltis assumes that (5.17) holds, or that – as in Kaldor's models – labour is fully employed. We have seen that growth equilibrium is possible without full employment in a Kaldor-type model, in which case (5.17) does not apply. Thus, even if some parametric change (say an increase in α) increases $-\hat{a}_0$, the higher rate of technological change will have no effect on the rate of growth of the economy in addition to that directly caused by the parametric change.[21]

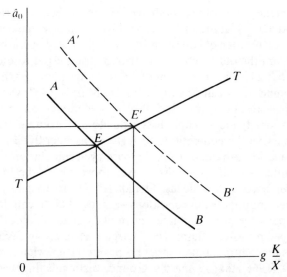

Figure 5.4 A model with Eltis's technical progress function

5.4 Technological change in Kalecki-Steindl models

We referred to three possible outcomes for the Kaldorian model, but discussed only two. The remaining, demand constrained case yields the same equilibrium as the Kalecki-Steindl model of chapter 2. Using Kaldor's technical progress function, we now examine the role technological progress plays in this model.

Recall from chapter 2 that in this model the rate of growth of capital depends on the parameters of the investment function (on α, β and τ in the linear case), s and z, and is independent of the input–output ratios. Also recall that the equilibrium value of the rate of profit, r^*, depends only on these parameters (and is independent of the technological parameters), and by (2.15), for given r and z, $\hat{K} = \hat{X}$.

The linear technical progress function (5.13′) can therefore be written as

$$\hat{X} - \hat{L} = T_0 + T_1 [\hat{X} - \hat{L}]$$

which implies (5.19). Thus the parameters of the technical progress function determine $-\hat{a}_0$, and whether or not a_1 changes is irrelevant.[22] Technological change cannot affect the rate of growth of output, which depends only on the parameters of the saving and investment functions and the markup rate. It follows that more rapid technological change reduces the rate of employment growth and thus increases unemployment.[23] The only way in which technological change can affect the rate of growth is by affecting the parameters of the investment and saving functions, or the markup rate.

There are reasons for believing that technological change will affect investment parameters. One of the central features of Schumpeter's (1934) writing is that a higher rate of technological change affects the rate of investment. Kalecki (1971) argues that technological change makes new machines more productive than old ones, raising the real costs of operating the latter. The profits from the old machines are thereby transferred to the new ones. This implies that a faster rate of technological change increases the rate of investment.[24] Similar ideas can be extracted from Marx's (1867) view that technological change results from, and leads to, competition between capitalists. As mentioned above, capitalists who do not accumulate and innovate will not survive the competitive struggle. Consequently, if greater possibilities of innovation are opened up, the pace of accumulation can be expected to accelerate. Not all economists – even those within the Marxist tradition – however accept these ideas. Steindl (1952), in his analysis of accumulation under monopolistic conditions, argues that 'technological innovations accompany the process of investment like a shadow, they do not act on it as a propelling force'. Baran (1957) argues strongly in favour of Steindl's view in his first edition, but changes his position somewhat later on. Baran and Sweezy (1966) take a middle ground, distinguishing between 'normal' and 'epoch-making' innovations. The former do not increase investment since monopoly capital tends to place restrictions on the implementation of the new methods and new products to protect existing capital values. But the latter (examples being the steam engine, the railroad and the automobile) tend to 'shake up the entire pattern of the economy and hence create vast investment outlets in addition to the capital which they directly absorb'.[25]

Assuming, for the sake of argument, that a higher rate of innovation does increase the parameters of the investment function, the rate of growth will increase in both the neo-Keynesian and Kalecki-Steindl models. In either case the effects on labour absorption can be found from

$$\hat{L} = g - \hat{a}_0$$

which is positive if α is very responsive to $-\hat{a}_0$ so that g increases more than $\hat{a}_0$ falls. In the Kalecki-Steindl model a faster rate of technological change will not change the distribution of income, which depends on z; in the neo-Keynesian model a higher rate of growth is possible only with a higher rate of profit and, given a_1, a lower wage share.

Technological change may have an effect on the rate of growth of the economy in the Kalecki-Steindl model by affecting other parameters as well. Kalecki (1941) suggests that a higher rate of technological change may raise the degree of monopoly power by increasing the scale of investments required, and this increases z. Sylos-Labini (1956) argues that a central feature of oligopolies is that technological change does not translate into lower prices but higher z. Finally, if faster technological progress reduces the rate of labour absorption and thus increases unemployment over time (adding to the reserve army),[26] as argued in chapter 4, this may increase z. As discussed in chapter 2, the rise in z will reduce the rate of growth (although qualifications have been noted in other chapters). If a higher rate of technological change also increases spending, the effect on the growth rate would depend on the strength of the two opposed forces. Thus a higher rate of technological change can provide an impetus to investment and yet be immiserizing in the sense of reducing the rate of growth of the economy. The Marxian falling rate of profit haunts us again, and this time without an increase in capital intensity.[27]

5.5 Conclusion

This chapter has introduced technological change in simple ways into our four basic models.

Our analysis suggests that growth and distribution can be affected by technological change in a variety of different ways, depending on the kind of model we examine, and the mechanisms through which it operates. Certain mechanisms will work only in certain types of models: for example, technological change will affect growth through its effect on the effective supply of labour in the neoclassical model, but not in others. For a particular model, the effect depends on the nature of the mechanisms: for example, in the Kalecki-Steindl model it depends on how exactly technological change affects the demand parameters and the markup. One thing that we can conclude is that there is no general presumption that technological change will always and everywhere improve growth and distribution.

CHAPTER 6

Two-sector models

6.0 Introduction

This chapter extends our basic models of growth and income distribution to allow for more than one sector. For simplicity we consider only two-sector models, but the analysis may be applied to more general cases.

The existence of more than one sector in the real world does not necessitate the construction of multi-sector models if our purpose is merely to examine the general issues of growth and income distribution. Otherwise, the aggregation implicit in all one-sector models would be inappropriate. Yet, one-sector models are not adequate in all instances and for all purposes. When there are important and systematic differences in the nature of different sectors and their role in the economy, one-sector models may provide inadequate formalizations of actual economies even for the understanding of what are generally considered macroeconomic issues. For instance, if there are some sectors which produce investment goods and some which produce consumption goods, or alternatively, if some sectors have flexible prices and others have fixed prices, one-sector models not taking these differences into account may provide incorrect results.[1] Even without systematic differences between sectors, some types of questions cannot be analysed with one-sector models. This chapter examines two-sector models to examine economies in which one sector produces consumption goods and the other investment goods, and to examine the determination of relative prices and issues relating to intersectoral capital mobility, which cannot be studied in a one-sector framework.

In dealing with two sectors, however, we shall abstract from issues considered in the last three chapters, including money, inflation, substitution in production, and technological change, and from open economy issues which will be considered later.

Section 6.1 will examine a general framework for the analysis of long-run equilibrium positions,[2] using the classical concept of intersectoral equalization of profit rates due to competition, and consider several models as 'alternative closures' of the framework. In addition to extensions of our basic models, we examine two new models, one combining features of two basic models to analyse the fixprice-flexprice distinction, and another which has received much attention in the consumption-investment goods literature on planning.

Section 6.2 will examine the dynamics of the economy out of long-run equilibrium, focusing mainly on the issue of intersectoral capital mobility in neo-Marxian and Kalecki-Steindl models.

6.1 Models of long-run equilibrium

6.1.1 The general framework

We assume that the economy is in long-run equilibrium in the classical sense. This involves the usual classical assumption that sectoral rates of profit are equalized due to competition, and that the economy is on a balanced-growth path with stocks of productive capital in each sector growing at the same rate.[3] The assumption that the economy is always in long-run equilibrium implies that we cannot examine the behaviour of the economy out of that long-run equilibrium, and cannot analyse whether there are forces that make it converge to that equilibrium. It should be made clear that by long-run equilibrium we do not imply that the economy *actually* converges to it, but that it is a centre of gravity around which the economy hovers, not straying very far from it due to forces pulling the economy towards it. The forces equalizing rates of profit will be discussed shortly, while disequilibrium issues are postponed to section 6.2.

The general framework we have chosen, following Hicks (1965, chapter XII), is as simple as we could make it for our purposes. We assume that there are two sectors of producton in the economy, producing one good each. One good is a consumption good and the other an investment good; there are no intermediates. Each sector has a Leontief (fixed coefficient and constant returns to scale) technology, producing output with the capital good and homogeneous labour, and there is no depreciation of the capital good. Labour is assumed to be perfectly mobile between the two sectors, so that the wage rate is equalized between them. The model can be extended to allow for more commodities, intermediate goods, and technological substitution.

These assumptions imply that we may write two quantity equations and two price equations for the economy, given by

$$X_1 = C(a_{01}X_1 + a_{02}X_2), \tag{6.1}$$

$$X_2 = g(K_1 + K_2), \tag{6.2}$$

$$P_1 = P_2(K_1/X_1)r + Wa_{01}, \tag{6.3}$$

$$P_2 = P_2(K_2/X_2)r + Wa_{02}, \tag{6.4}$$

where X_i denotes the level of output, P_i the price, K_i the stock of the capital good employed, and a_{0i} the technologicaly fixed labour-output ratio, all in sector i for $i = 1$ and 2, and the other symbols have the same meaning as before.

Note that $W, r,$ and g are the same for both sectors and the profit rate is defined by valuing inputs and outputs at the same prices.

As in previous chapters, we assume that workers do not save and capitalists save a constant fraction s of their income, so that

$$CL = (W/P_1)(a_{01}X_1 + a_{02}X_2) + (1-s)r(P_2/P_1)(K_1 + K_2), \qquad (6.5)$$

where $L = a_{01}X_1 + a_{02}X_2$, is total employment.

We may write (6.1) through (6.5) as

$$sr = g, \qquad (6.6)$$

$$X_1/K_1 = C[a_{01}(X_1/K_1) + a_{02}(X_2/K_2)/k], \qquad (6.7)$$

$$X_2/K_2 = g(k+1), \qquad (6.8)$$

$$1 = p(K_1/X_1)r + Va_{01}, \qquad (6.9)$$

$$p = p(K_2/X_2)r + Va_{02}, \qquad (6.10)$$

where $k = K_1/K_2, p = P_2/P_1,$ and $V = W/P_1$. Here (6.6) is derived from (6.1), (6.2), and (6.5), and (6.7) through (6.10) follow from (6.1) through (6.4), respectively. Note that V is the real wage, since good 1 is the only consumption good.

Three comments are in order for (6.6) through (6.10). First, they imply consumption-growth and wage-profit frontiers as in the one-sector case. This is shown by deriving the potential frontiers which set $K_i/X_i = a_{2i}$. Substituting these in (6.7) and (6.8) and eliminating k gives

$$(1 - Ca_{01})(1 - a_{22}g) - a_{02}a_{21}Cg = 0 \qquad (6.11)$$

which implies an inverse relation between C and g with

$$dC/dg = -[a_{22} + C(a_{02}a_{21} - a_{01}a_{22})]/[a_{01} + g(a_{02}a_{21} - a_{01}a_{22})].$$

This shows that the curvature of the potential consumption-growth frontier depends on $a_{02}a_{21} - a_{01}a_{22}$, the sign of which depends on the relative capital intensity of the two sectors. If the capital intensities are identical, the relation is linear; if the investment good is more (less) capital intensive than the consumption good sector, the frontier is concave (convex) to the origin. Similarly, substituting the full capacity conditions in (6.9) and (6.10) and eliminating p gives

$$(1 - Va_{01})(1 - a_{22}r) - a_{02}a_{21}Vr = 0 \qquad (6.12)$$

which implies an inverse relation between V and r which is exactly the same as that between C and g; this is the potential wage-profit frontier. We may analogously obtain frontiers for given levels of $K_i/X_i > a_{2i}$ showing excess capacity.

Secondly, they contain mainly accounting relations; the only *economic* assumptions they embody are those regarding consumption, the nature of technology, and long-run equilibrium. Thirdly, against five equations, we have eight variables, $V, p, X_1/K_1, X_2/K_2, r, g, C$, and k, so that to get a closed model we need three more equations.

We will first consider four different sets of three equations, giving us four alternative closure es, following the basic models of chapter 2, and they will be given the same r .mes as before. But other closures are possible. Some may use different assumptions for different sectors (as opposed to our use of symmetrical closures); we will examine one which synthesizes neo-Keynesian and Kalecki-Steindl features, similar to models recently used for analysing agriculture-non-agriculture interaction in less-developed economies. A final closure will introduce a planning agency which directs the allocation of investment.

6.1.2 The neoclassical model

The neoclassical closure assumes full capacity utilization, or

$$K_1/X_1 = a_{21},\tag{6.13}$$

$$K_2/X_2 = a_{22},\tag{6.14}$$

where a_{2j} is the technologically fixed capital-output ratio in sector $j = 1, 2$. The closure further assumes that in long-run equilibrium, the economy grows with full employment, so that[4]

$$g = n,\tag{6.15}$$

where, as before, n is the fixed rate of growth of labour supply.

The model is now closed. The equilibrium may be depicted as shown in Figure 6.1. As shown in (c), (6.15) determines g^* and (6.6) then solves for r^*. With (6.13) and (6.14) fixing $(K_i/X_i)^*$, we obtain the $V-r$ and $C-g$ curves of (a) and (d) given by (6.12) and (6.11);[5] these solve for V^* and C^*. As shown in (e), (6.7) solves for k^*. Finally, given r^* and V^*, p^* can be solved from (6.9) and (6.13) represented by the $P_1 P_1$ curve of (b) given by

$$V = (1/a_{01}) - (a_{21} r^*/a_{01})p$$

or from (6.10) and (6.14) represented by the OP_2 curve showing

$$V = [(1 - a_{22} r^*)/a_{02}]p$$

Note that these two curves must intersect at p^*, given (6.12).

The figure can be used to analyse the effects of parametric shifts. A rise in n, for example, will in (c) increase g^* and r^*, in (a) reduce V^*, in (d) reduce C^*, and in (e) reduce k^*. In (b) the increase in r^* rotates the OP_2 and $P_1 P_1$ curves downwards, confirming that V^* falls, and shows that the effect on p^* is ambiguous (depending on the relative capital intensity of the two sectors). Note that, as in the one-sector model, the higher g^* implies a lower C^*,

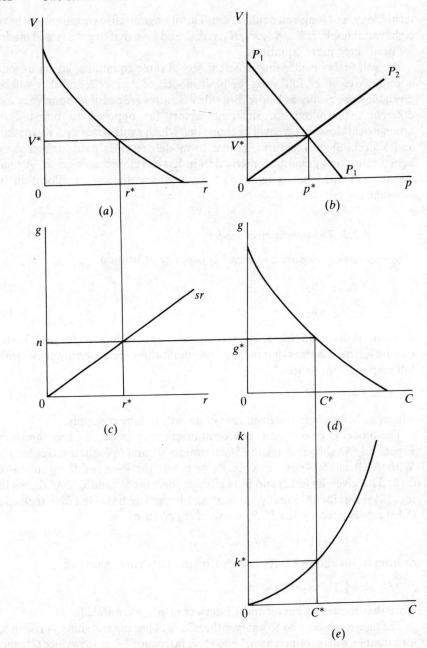

Figure 6.1   Long-run equilibrium in the two-sector neoclassical model

a higher r^* and a lower V^*; the higher g^* also requires a lower k^* or a higher relative size of the investment goods sector to allow the economy to accumulate capital faster.

Neoclassical two-sector models include those of Uzawa (1961, 1963), Solow (1961), Drandakis (1963), and Inada (1963). These models differ from ours by assuming variable production coefficients in each industry, and sometimes by making different savings assumptions; they also examine long-run dynamics and stability, which we discuss later.

6.1.3 The neo-Marxian model

The neo-Marxian closure also assumes full capacity utilization, so that (6.13) and (6.14) are satisfied. It further assumes that the real wage is fixed at 'subsistence', so that

$$V = \tilde{V}. \tag{6.16}$$

Equilibrium can be shown in Figure 6.2, where the diagrams are the same as in Figure 6.1, except that in (d) (which plays the same role as (b) in the previous figure) $P_1 P_1$ and OP_2 represent (6.8) and (6.9), after substituting (6.13), (6.14), and (6.16), and are given by the equations

$$p = (1 - \tilde{V} a_{01})(1/a_{21} r)$$

and

$$p = a_{02} \tilde{V}[1/(1 - a_{22} r)]$$

respectively. An increase in $\tilde{V}$ will in (a) reduce r^*, in (b) reduce g^*, in (c) increase c^* and in (e) increase k^*. The $P_1 P_1$ curve will be pushed down and $P_2 P_2$ pushed up, confirming the reduction in r^* but showing that the effect on p^* is again ambiguous.

Findlay (1963) attributes this model to Robinson (1956), but her model (see especially Robinson, 1962), in which the desired accumulation plays a central role, is closer to the neo-Keynesian closure examined next. The crucial feature of this model is the Marxian feature of the fixed real wage,[6] and it conforms to models available in Morishima and Catephores (1978), and Marglin (1984a), although Marglin's model has many sectors and only circulating capital. It is also the same as Vianello's model (1985), although the latter uses a Sraffian closure with an exogenous r rather than V. The model differs from some Marxian multi-sector models in not examining the dynamics of intersectoral capital flows (see section 6.2.2), not using value categories (Harris, 1978), assuming away the problem of realization (Foley, 1986), and ignoring the impact of labour supply on the real wage (Harris, 1978). Some of these extensions can be made following the discussion in chapter 3, dynamics will be considered later, and effective demand will appear in the neo-Keynesian and Kalecki-Steindl closures.

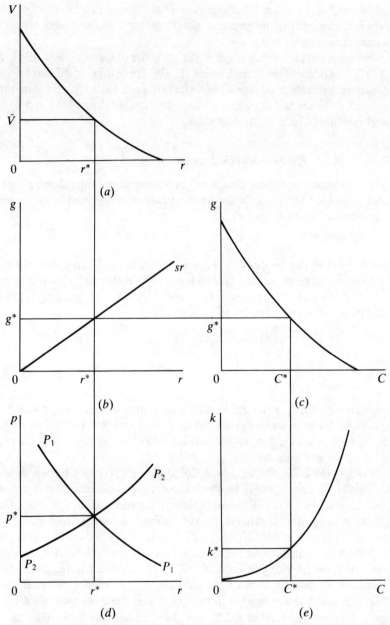

Figure 6.2   Long-run equilibrium in the two-sector neo-Marxian model

The balanced-growth interpretation of the Marxian model has been criticized. Sardoni (1981), for instance, argues that Marx's conditions for expanded reproduction, require only positive, and not equal, rates of sectoral growth; moreover, Marx does not use expanded reproduction models to depict actual economies, but to show under what conditions crises do not occur. While this may be true, the value of balanced-growth analysis – even in a Marxian context – is that it shows the movement of the economy in a simple way, using the tools of equilibrium analysis.[7] This can then be used as a basis for studying the effects of parametric changes which make the economy diverge from the balanced-growth path (since such changes in general alter k^*, and hence imply divergent rates of sectoral growth). Further, the analysis of equilibrium can pave the way for the analysis of disequilibrium, as we shall see in section 6.2.

6.1.4 The neo-Keynesian model

The neo-Keynesian closure also assumes full capacity utilization in each sector, so that (6.13) and (6.14) hold. The model is then closed with the desired accumulation function

$$g = g(r), \tag{6.17}$$

where $g'(r) > 0$.

The determination of equilibrium in this model can be shown in Figure 6.3 where all diagrams but (c) are the same as that in Figure 6.1. In (c) the GG curve showing (6.17) and the sr line showing (6.6) determine r^* and g^* and the other variables are then solved as in Figure 6.1.

This model corresponds to the work of Robinson (1956, 1962), and also to Marglin's (1984a) neo-Keynesian closure.

6.1.5 The Kalecki-Steindl model

The crucial feature of the Kalecki-Steindl closure is the existence of market power of firms in each industry. We formalize this with markup pricing equations as before (although now we need one for each sector), given by

$$P_1 = (1 + z_1)Wa_{01}, \tag{6.18}$$

$$P_2 = (1 + z_2)Wa_{02}, \tag{6.19}$$

where z_i is the fixed markup rate in sector i.[8] As discussed in chapter 3, the markup rates depend on factors such as industrial structure and the state of the class struggle. In a more general model, with intermediate goods, both z_i's will be affected by the wage bargaining process; in this model, since only z_1 affects the real wage, z_2 may not be independent of it. Both will be affected by concentration rates; since these can differ between sectors, there is no reason for the markups in the two sectors to be equal. As in the one-sector version,

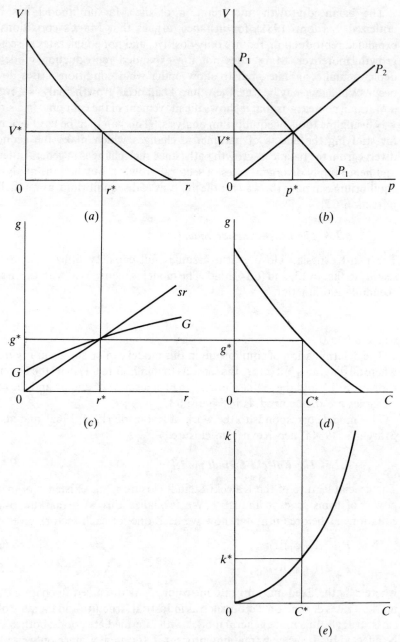

Figure   6.3   Long-run equilibrium in the two-sector neo-Keynesian model

these equations embody a specific *theory* of price formation, in which large firms set prices as a markup on prime costs, and change the levels of output to meet demand; this is possible because of the existence of excess capacity. We assume that the z_i's are given at a level sufficient (given the other parameters) to ensure equilibrium with excess capacity.

The third equation is the desired accumulation function,

$$g = g(r, X_1/K_1, X_2/K_2), \tag{6.20}$$

where all three partial derivatives are positive. This should properly be thought of as a reduced-form equation, which shows how investment plans are made in equilibrium, rather than as a behavioural equation. The behavioural equations on which it is based are not discussed formally here,[9] but a plausible mechanism yielding them may be outlined as follows. The desired accumulation rate of firms in each sector depends on the sectoral rate of profit, the interest rate, and the degree of sectoral capacity utilization, due to neo-Keynesian and Kaleckian, standard neoclassical, and Steindlian reasons, respectively. In the short run the rates of profit may diverge from the common interest rate (established by the mobility of financial capital), and hence from each other. In long-run equilibrium, where the sectors have the same rate of growth of capital stock, the rates of profit are equalized with the interest rate, and hence with each other, so that the growth rate of capital stock depends on the common rate of profit and capacity utilization rates. Specific forms of behavioural functions may have to be postulated to obtain this reduced form.[10]

The model may be solved as follows. From (6.18) we get

$$V^* = 1/[a_{01}(1+z_1)], \tag{6.21}$$

while (6.18) and (6.19) imply

$$p^* = [a_{02}(1+z_2)]/[a_{01}(1+z_1)]. \tag{6.22}$$

Substitution of (6.18) and (6.19) into (6.9) and (6.10) shows

$$r = [z_1/(1+z_2)](a_{01}/a_{02})(X_1/K_1) \tag{6.23}$$

and

$$r = [z_2/(1+z_2)](X_2/K_2) \tag{6.24}$$

which imply

$$X_1/K_1 = (z_2 a_{02}/z_1 a_{01})(X_2/K_2). \tag{6.25}$$

Now substituting (6.20) in (6.6), and for r and X_1/K_1 from (6.24) and (6.25), respectively, we obtain

$$s[z_2/(1+z_2)]X_2/K_2$$
$$= g((z_2/(1+z_2)X_2/K_2(z_2 a_{02}/z_1 a_{01})X_2/K_2, X_2/K_2) \tag{6.26}$$

which solves for $(X_2/K_2)^*$. The equilibrium is shown in Figure 6.4, where the OS line shows the right-hand side of (6.26) and the GG line shows the left-hand side; it is assumed that the latter is flatter than the former. Notice that the GG curve is rising because a higher X_2/K_2 makes all three arguments of the g function increase, all of them increasing g. The figure shows how $(X_2/K_2)^*$ and g^* are solved in (6.26). Substituting these values into (6.24) and (6.25) solves for r^* and $(X_1/K_1)^*$, and (6.8) then solves for k^* and (6.7) for C^*.

As an example of a parametric shift, consider a fall in z_1. By (6.21), this increases V^*. From (6.26) and Figure 6.4 we see that the GG curve shifts up, raising g^* and $(X_2/K_2)^*$; for a given s this implies a higher r^*. The result derived for the one-sector case (see chapter 2) holds in this two-sector case, and for the same reason; the fall in the degree of monopoly power in sector 1 raises the real wage, changes the distribution of income from high-saving capitalists to workers, increases consumption, raises aggregate demand, reduces excess capacities, and thereby raises both profit and growth rates.

Finally, we consider whether equilibrium of the type we are considering (with equalized profit rates) is at all possible with excess capacity. In section 3.4 we examined this question in a one-sector setting, and we now examine whether, due to the fact that we are in a multi-sector environment, any additional problems arise. In a multi-sector setting, we are dealing with long-run equilibrium positions in which classical competition equalizes rates of profit intersectorally. Eatwell (1983) states that in the classical long-run

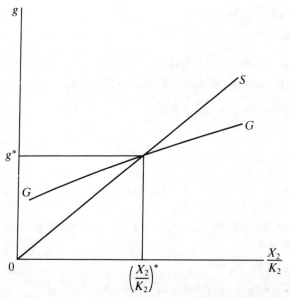

Figure 6.4   Long-run equilibrium in the two-sector Kalecki-Steindl model

equilibrium full capacity will exist, and Vianello (1985), although considering short-run deviations informally, assumes that in the long run 'normal' capacity utilization will prevail. However, they present no mechanism which would bring this about. While markup rates may change due to excess capacity and this *may* take the economy to full capacity utilization, the issue here is whether this mechanism is in the same time horizon in which classical competition works. Markups *may* change if capital flows involve the movement of firms between sectors and this changes the degree of monopoly power in different sectors; but since mobility in classical competition essentially refers to the mobility of financial capital it is by no means obvious that such changes will occur. Ciccone (1986) argues that classical competition ensures that the relative sizes of industries is such that investment has the same profitability in all industries, and has nothing to do with aggregate demand and capacity utilization. This is shown in our model where, with rates of profits equalized and with excess capacities prevailing, there is no mechanism by which further capital movements can achieve full capacity.[11]

6.1.6 A fixprice-flexprice model

We now consider a model combining neo-Keynesian and Kalecki-Steindl elements. We assume that the investment good industry operates with excess capacity, so that P_2 is given by (6.19), while the consumer good industry operates with full capacity, so that (6.13) is satisfied. Thus sector 1 is flexprice, the price varying to clear the market, while sector 2 is fixprice, with capacity utilization clearing the market at a price fixed by costs and the markup.

Investment can be assumed to depend on the rate of profit and the degree of capacity utilization in sector 2 (since full capacity utilization prevails in sector 1), so that

$$g = g(r, X_2/K_2). \tag{6.27}$$

From (6.19) we get (6.24) which, using (6.6) and (6.27), yields

$$sr = g(r, r(1 + z_2)/z_2) \tag{6.28}$$

which solves for r^* and g^*. Given r^*, (6.24) determines $(X_2/K_2)^*$. Substituting these and (6.13) into (6.9) and (6.10) solves for p^* and V^*. The rest of the variables can be determined as in earlier models. A fall in z_2 implies, from (6.28) (and assuming stability), a higher r^* and g^*, and (6.24) a higher $(X_2/K_2)^*$.

This model resembles the fixprice-flexprice models (see Cardoso, 1981, Taylor, 1982, 1983, Thirlwall, 1986) used in the analysis of agriculture-industry interaction for less-developed economies, where following Kalecki (1971) and Hicks (1974), the agricultural sector is assumed to be flexprice and the non-agricultural sector, fixprice. The only major difference between ours and these is that they assume that the fixprice good is consumed and invested, while ours only allows it to be invested. The models also examine the nature of short-run equilibrium, and examine intersectoral capital mobility in response to short-run profit-rate differentials (see Taylor, 1983); in long-run equilibrium growth rates and rates of profit are equalized. The models usually take the rate of growth of agriculture to be given, make some assumption regarding capital mobility between the two sectors, and show that the long-run equilibria are stable. They can easily be generalized in the way discussed here to allow the rate of growth to be endogenous. Hicks (1974) argues that in flexprice markets traders carry stocks to gain from price changes; stock, rather than flow, equilibrium is the appropriate concept. Our model ignores the holding of stocks of the flexprice good, and analyses flow equilibrium with demand and supply coming from consumption and production, respectively. But Dutt (1986b) develops a short-run agriculture-industry model where the flexprice sector has stock equilibrium.

6.1.7 The Feldman-Mahalanobis-Domar model

No discussion of models with consumption and investment goods sectors is complete without the Feldman-Mahalanobis-Domar (FMD) model (see Mahalanobis, 1953, Domar, 1957), which was used to justify the emphasis on investment goods sectors in the early Soviet and Indian plans. While Feldman initially applied his model to the problem of investment allocation by state planners in the socialist Soviet economy, the model has been applied to predominantly capitalist ones, where the government can determine investment allocation (at least in the long run) through tax-subsidy schemes. It is thus of some relevance to examine the model in terms of our two-sector framework.[12]

The model assumes full capacity utilization in both sectors, so that (6.13) and (6.14) hold. The planners are assumed to allocate – in the long run – a fixed share of investment to each sector; Θ is the share allocated to the investment goods sector.

Since (6.14) holds, the output of the investment goods sector, $X_2 = K_2/a_{22}$. Since Θ of this is invested in sector 2,

$$g^2 = \Theta/a_{22} \tag{6.29}$$

where g^i is the rate of accumulation in sector i, and

$$g^1 = [(1 - \Theta)/a_{22}]/k. \tag{6.30}$$

In steady state, which is all that we are concerned with now, $g^1 = g^2$, so that (6.29) and (6.30) imply

$$k = (1 - \Theta)/\Theta \tag{6.31}$$

which shows Θ fixes k^*. Once k^* is solved, (6.8) determines g^*, (6.6) determines r^*, and substitution into (6.9) and (6.10), with the full capacity utilization equations, solves for V^* and p^*.

Note that, if the planners increase Θ, $k*$ and $g*$ will increase, while $V*$ will decrease. In the capitalist economy we have considered (with full capacity utilization and equalized rates of profit), expansion of the investment good sector can increase the economy's rate of growth but only by worsening the distribution of income.

6.2 Dynamics out of long-run equilibrium

We now examine some of the models to analyse dynamics when the economies are not in long-run equilibrium as defined in section 6.1, that is, not on their balanced growth path. Neoclassical models typically examine dynamics out of long-run equilibrium by examining the evolution of sectoral rates of growth over time, though maintaining the assumption of equalized rates of profit (or rental) in the short run; we briefly summarize the main implications of these models, and examine a somewhat different formulation, in section 6.2.1. These models, however, do not examine the implications of differential rates of profits, which seem to be at the heart of the classical dynamics of competition; to examine this issue we introduce differential profit rates in the neo-Marxian model in section 6.2.2. While this model implies that the economy ends up in long-run equilibrium with rates of profit intersectorally equalized, this result does not hold for all models: section 6.2.3 considers dynamics in a Kalecki-Steindl model with monopoly power, and section 6.2.4 analyses dynamics in the FMD model with government intervention, to illustrate this.

6.2.1 Neoclassical dynamics

As noted earlier, standard neoclassical two-sector models (Uzawa, 1961, 1963, Solow, 1961, and Drandakis, 1963) differ from ours in allowing for factor substitution. They are thus able to consider dynamics out of long-run equilibrium in a way not possible in ours. They distinguish between a short-run (or temporary) equilibrium and a long-run (steady-state) equilibrium. In the former, total stocks of capital and labour are given, all markets (including those for labour and capital) clear due to price change, and all factors (including capital) are perfectly mobile so that the wage and rental (profit) rates are equalized between sectors (and to the relevant marginal products). In long-run equilibrium, in addition to all the conditions of temporary equilibrium being fulfilled, the two sectors grow at the same rate and total capital grows at the same (exogenously) fixed rate as labour.

A sufficient condition for the stability of this long-run equilibrium (when factor substitution is allowed) is the equality of capital-labour ratios in the two sectors at all factor-price ratios. This condition in effect reduces the model to a one-sector model. If the capital-labour ratios are different, a sufficient condition for stability, which is called the capital-intensity condition, is that the capital-labour ratio (at all wage-profit ratios) for sector 1 (the consumption good sector) is greater than that for sector 2. This proposition has been found to be true under a variety of assumptions about saving behaviour, including identical saving propensities out of profits and wages (Uzawa, 1963), no savings from wages and no consumption from profits (Uzawa, 1963), and different saving propensities, with a higher one out of profits (Drandakis, 1963); it is also valid for our case with no saving out of wages and a constant saving propensity out of profits. The condition, however, is not a necessary one. If it is not satisfied and if the saving propensity out of profits is higher than that out of wages, a sufficient condition for stability is that the elasticities of substitution in production in the two sectors exceeds unity. If none of these sufficient conditions is satisfied, the models cannot guarantee stability, although examples of stability can still be found.

While these dynamics with smooth production functions are most popular in the neoclassical literature, there are others which can generate the same type of long-run equilibrium. As an example, consider an extension of the one-sector neo-Marxian model of section 3.2.2, where we assume that the real wage is given in the short run, the rates of profit between sectors are equalized, and the two sectors grow at the same rate as well.[13] Over time V changes according to

$$dV/dt = \Omega[g - n].$$

We stay closer to neoclassical rather than Marxian assumptions by taking the saving propensity out of profits and wages to be the same. This implies, due to the saving-investment equality, that

$$\sigma(P_1 X_1 + P_2 X_2) = P_2 X_2 \tag{6.32}$$

which can be rewritten, using (6.13) and (6.14), as

$$\sigma\{[(1/a_{21})k/p] + (1/a_{22})\} = 1/a_{22}. \tag{6.33}$$

To examine the nature of the dynamics in this model we explore the relation between g and V, by examining those between V and p, between p and between k and g, in turn. The last two relations follow from (6.33), and (6.8), and (6.14), respectively; they imply $dp/dk > 0$ and $dg/dk < 0$. The first can be deduced from (6.9) and (6.10) after substituting (6.13) and (6.14): this gives

$$p = (a_{22}/a_{21}) + V[a_{02} - (a_{22}a_{01}/a_{21})] \tag{6.34}$$

which implies that dp/dV has the same sign as $[a_{02}/a_{22} - a_{01}/a_{21}]$. If the capital-intensity condition is satisfied, we have $dp/dV > 0$, so that, given the signs of dg/dk and dk/dp, we can conclude that $dV/dg < 0$. Figure 3.5 can again be used to show this relation as curve gg. Given n, the long-run equilibrium is at V^e and the adjustment process is stable. But, if the capital-intensity condition is violated, the gg curve will have a positive slope, and the adjustment process will be unstable.[14] We therefore find that stability in this model also requires the capital-intensity condition.[15]

If we assume that workers consume all their income and capitalists save a fraction s of theirs, we must replace (6.33) by

$$sr(k+1) = 1/a_{22} \qquad (6.35)$$

which follows from (6.6) and (6.8). If we now use (6.9), (6.10), (6.13), (6.14), and (6.35) we get

$$k = [(1-s)(1 - Va_{01})a_{22} + Va_{02}a_{12}]/s(1 - Va_{01})a_{22} \qquad (6.36)$$

which implies

$$dk/dV = a_{21}a_{02}/(1 - Va_{01})_2 sa_{22} \qquad (6.37)$$

which is positive if profits are positive. From (6.8) and (6.14) we get $dg/dk < 0$ as before, so that in this case we have $dg/dV < 0$. Figure 3.5 can again be used to depict this case, and the adjustment process is necessarily stable, without requiring the capital-intensity condition.[16] Note that under the same assumptions about saving behaviour and (zero) elasticities of substitution in production, the stability of the standard neoclassical adjustment process requires the capital-intensity condition (as a sufficient condition).

Before we leave the neoclassical adjustment story, it should be pointed out that it does not consider movement of capital between sectors over time in response to rate of profit differentials. Even in short-run equilibrium, completely malleable physical capital can go from one sector to another so as to equalize the profit rate between sectors. As we shall see, this is a process quite different from the classical convergence process.[17]

6.2.2 Classical and neo-Marxian dynamics

The classical economists take competition to be a process which equalizes rates of profit in different sectors. Sraffa (1960) (among others) formalizes this notion of competition using prices (which, following Marx, can be called the prices of production) which equalize the rates of profit, and interprets the position as a centre of gravitation. However, he does not analyse formally the process, so important to the classical economists, by which this equalization comes about. Several attempts have been made recently to formalize the classical dynamics and examine the question of stability of the centre of gravity. These assume that prices respond positively to excess demand gaps, while profit rate differentials determine the relative growth rates of production or capital in different sectors (those sectors have higher profits growing faster). They find that it is not generally true that the economy will converge to the equilibrium with equilized profit rates, under these conditions.

It has been found that a condition for the stability of the prices of production in a two-sector framework is that the organic composition of the capital in the consumption goods sector is greater than that in the investment goods sector. Nikaido (1983), who initially derived this condition for a Marxian model of simple reproduction, writes that:

> the non-equalization of profit rates always prevails if the capital good sector has a higher organic composition, the situation which is often presumed in the Marxist view. Thus the result of our examination seems to suggest that prices of production are not natural prices to which market prices tend to be attracted, but just normative constructons, an attribute of the social relations of production, like labor values.

Allowance for positive saving in a model of expanded reproduction makes Nikaido (1985) modify his conclusion somewhat; with very high or very low savings rates (of capitalists) stability can be attained, but for intermediate values of that parameter, the capital-intensity condition is usually (sufficiently low investment response parameters are also helpful) again required.[18]

Several subsequent contributions try to introduce stability into Marx-Sraffa systems by modifying the underlying dynamics of the convergence process. Thus Flaschel and Semmler (1986b) assume that capitalists do not just look at profit rates, but also the direction of change in them, in determining the relative growth rates of different sectors. Franke (1986) and Dumenil and Levy (1986) introduce stability by allowing for inventory adjustment. Boggio (1986) assumes that prices do not directly respond to excess demand, thus departing from an essential feature of the classical mechanism. Finally, Boggio (1985) and Hosoda (1985) introduce strong price substitution effects in capitalist consumption to increase the changes of stability.[19] While these and other stabilizing influences can be introduced into the analysis, it is not quite clear to what extent they are important enough in reality to stabilize the model, or to what extent they do not deviate too far from the dynamics envisaged by the classical writers. It is therefore worth returning to the basic Nikaido-type models which do not introduce such modifications.

The importance of reconsidering the role of the capital-intensity condition arises because Nikaido's stability assumption involving organic compositions of capital can be shown to be almost identical to the capital-intensity condition of the neoclassical growth models mentioned above.[20] Non-neoclassical writers criticize neoclassical models for requiring this (and the other) condition(s). Harris (1978) writes that '[t]hese are obviously very special conditions, chosen on an ad hoc basis, without adequate theoretical justification. Outside of these very special conditions, no such relation as that required by the neoclassical parable can be found.' Those who agree with Harris should be embarrassed by the stability conditions that have been discovered for the Marx-Sraffa models.[21]

We consider the same economy as in section 6.1, but do not assume that rates of profit or growth rates are always equal between sectors.[22] Note that we are assuming that wages are paid after production, so that the wage fund is not a part of the capital on which the rate of profit is earned, and this sets our model apart from all of the other models which have analysed the stability condition, which make the fund a part of capital. Our model also considers fixed capital, while most of the others allow only for circulation capital.

Our assumptions imply that we must rewrite (6.9) and (6.10) as

$$1 = pa_{21}r_1 + Va_{01}, \tag{6.38}$$

$$p = pa_{22}r_2 + Va_{02}, \tag{6.39}$$

where r_i is the rate of profit in sector i.

Our saving assumptions imply that the demand for each good (in real terms) D_i, is

$$D_1 = V(a_{01}X_1 + a_{02}X_2) + (1-s)p(r_1K_1 + K_2r_2), \tag{6.40}$$

$$D_2 = s(r_1K_1 + r_2K_2). \tag{6.41}$$

As in neo-Marxian models above we assume that V is fixed, and there is an army of the unemployed, so that employment is determined by the demand for workers. V is assumed to be such as to yield positive profits in sector 1, so that $1 - Va_{01} > 0$. The rate of profit need not be the same in both sectors, and, if there is a divergence, capitalists will invest more in the sector which provides a higher rate of profit. This idea may be formalized by assuming

$$g^i = g^j + \mu(r_i - r_j) \tag{6.42}$$

for $i = 1, 2, j = 1, 2$, and $i \neq j$, where g^i is the desired rate of capital accumulation (investment to capital-stock ratio) in sector i and $\mu > 0$ measures the responsiveness of investment differentials to profit rate differentials.[23]

In the short run, this economy has given stocks of capital in each sector, so that output levels are fixed and markets clear through price variations. We assume that in short-run equilibrium demand equals output in each sector, so that desired and actual accumulation rates are equal.[24] In the long run, stocks of capital change according to sectoral rates of investment. We consider the short-run and long-run behaviour of the economy in turn.

We assume that, if the demand and supply are not balanced in the markets, the relative price level will adjust in the short run. In particular, we assume a short-run adjustment equation of the form

$$dp/dt = \Omega(D_2 - X_2), \tag{6.43}$$

where $\Omega > 0$.

In short-run equilibrium, $dp/dt = 0$, or $D_2 = X_2$. Using (6.41) this implies

$$X_2 = s(r_1K_1 + r_2K_2). \tag{6.44}$$

Note that this implies, from (6.38) and (6.39), that

$$X_1 = V(a_{01}X_1 + a_{02}) + (1-s)p(r_1K_1 + r_2K_2) \tag{6.45}$$

which implies $D_1 = X_1$, or demand-supply balance for sector 1 as well, proving

Walras's law for this economy. We will confine attention only to the sector 2 demand-supply balance.

From (6.38), (6.39), and (6.44) the short-run equilibrium level of the relative price is seen to be

$$p = s[X_1 - Va_{01}X_1 - Va_{02}X_2]/(1-s)X_2 \tag{6.46}$$

which is positive as long as

$$(1 - Va_{01})a_{21}/Va_{02}a_{22} > k. \tag{6.47}$$

Note that this condition requires that V is low enough to yield a positive profit in sector 1, so that $1 > Va_{01}$, which we have already assumed. It is equivalent to the condition that V is not high enough to make the consumption demand by workers exceed the output of the consumption good.

The short-run equilibrium values of the rates of profit can be determined as follows. From (6.38) and (6.39) we get

$$Va_{02}a_{21}r_1 + (1 - Va_{01})a_{22}r_2 = 1 - Va_{01}. \tag{6.48}$$

Dividing (6.44) by K_2 on both sides we get

$$skr_1 + sr_2 = 1/a_{22}. \tag{6.49}$$

From (6.48) and (6.49) we may solve for

$$r_1 = (1-s)(1 - Va_{01})/D, \tag{6.50}$$

$$r_2[sk(1 - Va_{01}) - Va_{02}a_{21}/a_{22}]/D, \tag{6.51}$$

where

$$D = s[(1 - Va_{01})a_{22}k - Va_{02}a_{21}].$$

Note that (6.47) ensures positive values for these variables.

Since (6.38), (6.39), and (6.44) imply that

$$D_2 - X_2 = s\{[X_1 - Va_{01}X_1 - Va_{02}X_2]/p\} - (1-s)X_2 \tag{6.52}$$

it is clear that the process of adjustment described by (6.43) is stable if the expression in square brackets in (6.52) is positive at short-run equilibrium. Equation (6.47) guarantees that it is.

In the long run sectoral capitals change according to (6.42). To examine these dynamics note that, since $\hat{k} = g^1 - g^2$, (6.42) implies

$$\hat{k} = \mu(r_1 - r_2). \tag{6.53}$$

Substitution from (6.50) and (6.51) into (6.53) implies

$$\hat{k} = \mu\{[(1-s)(1 - Va_{01}) - sk(1 - Va_{01}) + Va_{02}a_{01}/a_{22}]/D. \tag{6.54}$$

Long-run equilibrium is attained at

$$k^* = [(1-s)(1 - Va_{01})a_{22} + Va_{02}a_{21}]/s(1 - Va_{01})a_{22}. \tag{6.55}$$

At long-run equilibrium, since k is constant, $g^1 = g^2$, so that, from (6.42), $r_1 = r_2$, showing that the rates of profit are equalized at long-run equilibrium with balanced growth. Since an increase in k reduces the right-hand side of (5.54), this equilibrium is also seen to be stable.

We have thus proved that the economy modelled here will converge to a balanced-growth path, with equalized rates of profit, from arbitrary initial conditions.[25] In particular, long-run stability in this sense does not require the Nikaido capital-intensity condition required in other models. The difference, obviously, has to do with differences in assumptions. We have assumed that wages are not considered to be a part of capital and that capital is fixed capital, whereas in all models of classical competition they have been assumed to be a part of capital which is assumed to be circulating. While we do not wish to claim that our assumption is better,[26] the rising importance of machinery and equipment in modern industry suggests that capital today is more appropriately thought of as fixed capital, rather than wages fund or circulating capital.[27]

In any case, our result does show that the capital-intensity condition is not required in general, but only in specific models. We should stress that we have not proved that the classical competitive process is necessarily stable. Our particular two-sector model does not allow the types of problems that have been emphasized by Steedman (1984) in a many-commodity model with intermediate goods, but that does not mean that they cannot arise in the real world. We agree with Steedman that the moral of the story

> is *neither* that the classical competitive process converges towards the 'natural' configuration ... *nor* that it fails to do so. The moral is simply that there is a genuine question at stake here, whose answer is not self-evident. (Emphasis in original.)

We only claim that Nikaido was too hasty in dismissing the prices of production concept on the basis of a two-sector neo-Marxian model.

6.2.3 Dynamics with monopoly power

Formalizations of classical competition do not usually make clear the exact assumptions they embody about firm behaviour, but the postulate that prices adjust to excess demands usually is interpreted to require atomistic behaviour. Thus, in an economy with large firms, the dynamics may well be different. We now examine dynamics for an economy with firms possessing monopoly power, using the assumptions of the Kalecki-Steindl model.[28]

The economy we consider has the same basic characteristics as the two-sector models we have just considered, except that the real wage is not considered to be given, and excess capacity is allowed to exist. Output in each of the two sectors is produced by large firms which are sufficiently homogeneous within each sector to be considered as two representative firms. The economy can thus be represented by the following set of four equations which are the commodity balance and price equations for the two sectors:

$$X_1 = C(a_{01}X_1 + a_{02}X_2), \tag{6.56}$$

$$X_2 = g^1 K_1 + g^2 K_2, \tag{6.57}$$

$$P_1 = P_2(K_1/X_1)r_1 + Wa_{01}, \tag{6.58}$$

$$P_2 = P_2(K_2/X_2)r_2 + Wa_{02}. \tag{6.59}$$

As before, our savings assumptions imply (6.44). Firms with monopoly power set prices according to (6.18) and (6.19). Finally, we assume linear sectoral investment functions of the form used in the one-sector Kalecki-Steindl model, given by

$$g^1 = \alpha_1 + \beta_1 r_1 + \tau_1(X_1/K_1), \tag{6.60}$$

$$g^2 = \alpha_2 + \beta_2 r_2 + \tau_2(X_2/K_2). \tag{6.61}$$

Note, for later use, that (6.18) and (6.19) imply (6.21) and (6.22), which already solve for V^* and p^* in terms of the parameters of the model. Also, (6.58), (6.59), (6.18), and (6.19) imply

$$r_1 = [z_1/(1 + z_2)](a_{01}/a_{02})(X_1/K_1), \tag{6.62}$$

$$r_2 = [z_2/(1 + z_2)](X_2/K_2). \tag{6.63}$$

In the short run we assume that the stock of capital in each sector is given.

As already assumed, firms adjust output in response to demand in each sector. Assuming linear adjustment equation with adjustment coefficients set equal to unity for simplicity,

$$dX_1/dt = V(a_{01}X_1 + a_{02}X_2) + (1-s)p(r_1 K_1 + r_2 K_2) - X_1,$$

$$dX_2/dt = g^1 K_1 + g^2 K_2 - X_2.$$

At short-run equilibrium, $dX_i/dt = 0$, which implies

$$X_1 = V(a_{01}X_1 + a_{02}X_2) + (1-s)p(r_1 K_1 + r_2 K_2), \tag{6.64}$$

$$X_2 = g^1 K_1 + g^2 K_2. \tag{6.65}$$

Substituting from (6.21), (6.22), and (6.60) through (6.63) we can write these as

$$\begin{bmatrix} -sz_1/(1+z_1) & (a_{02}/a_{01})[1+(1-s)z_2]/(1+z_1) \\ \delta_1 & -(1-\delta_2) \end{bmatrix} \begin{bmatrix} X_1 \\ X_2 \end{bmatrix} = \begin{bmatrix} 0 \\ \alpha_1 K_1 + \alpha_2 K_2 \end{bmatrix}, \tag{6.66}$$

where $\delta_1 = \beta_1[z_1/(1+z_2)](a_{01}/a_{02}) + \tau_1$ and $\delta_2 = \beta_2[z_2/(1+z_2)] + \tau_2$. We can consequently solve for X_i/K_i as

$$X_1/K_1 = [1 + (1-s)z_2](1 + z_2)(a_{01}/a_{02})[\alpha_1 + (\alpha_2/k)]/z_1 D, \tag{6.67}$$

$$X_2/K_2 = s(1 + z_2)(\alpha_1 k + \alpha_2)/D, \tag{6.68}$$

where[29]

$$D = s[1 + z_2)(1 - \tau_2) - \beta_2 z_2] - [1 + z_2(1-s)][\beta_1 + \tau_1((1+z_2)/z_1)(a_{02}/a_{01}).$$

For meaningful values of X_i/K_i we require that the parameters of the investment functions, β_i and τ_i are sufficiently small, given s, a_{01} and z_i to satisfy

$$D > 0 \qquad (6.69)$$

and that the parameters of the model satisfy the restrictions $X_i/K_i < 1/a_{2i}$, where X_i/K_i are given by (6.67) and (6.68). Once the equilibrium values of X_i/K_i are determined we can show that the equilibrium values of the rates of profit are given by

$$r_1 = \{[1 + z_2(1-s)]\alpha_1/D\} + \{\alpha_2[1 + z_2(1-s)]/D\}(1/k), \qquad (6.70)$$

$$r_2 = (\alpha_s s z_2/D) + (\alpha_1 s z_2/D)k. \qquad (6.71)$$

The (local) stability of this short-run equilibrium requires that the Jacobian of the dynamic system, that is the matrix in (6.66), has a negative trace and a positive determinant. Since the trace is given by $-(s\{z_1/(1+z_1) + [1-\beta_2 (z_2/(1+z_2)) - \tau_2]\}$ and the determinant is $\det = [z_1/(1+z_2)]D$, where D has already been defined, the system is seen to be locally stable assuming (6.69).

In the long run, K_1 and K_2 change over time due to capital accumulation by firms. From the fact that $\hat{k} = g^1 - g^2$, (6.60) and (6.61) we have

$$\hat{k} = (\alpha_1 - \alpha_2) + [\beta_1 + \tau_1(a_{02}/a_{01})(1 + z_2)/z_1]r_1 - [\beta_2 + \tau_2(1+z_2)/z_2]r_2. \qquad (6.72)$$

For long-run equilibrium, $dk/dt = 0$. Substitution of (6.70) and (6.71) into (6.72) shows that the right-hand side of (6.72) falls with k, so that the equilibrium is stable.

It is, however, more instructive to consider these long-run dynamics graphically, as in Figure 6.5. The northwest and southeast quadrants, respectively, depict (6.71) and (6.70); the southwest quadrant plots the rectangular hyperbola given by $(1/k)k = k$. These three curves are used to derive the curve SS of the northeast quadrant, which thus shows combinations of r_1 and r_2 which are consistent with short-run equilibrium (market clearing for both sectors) at different levels of k. Setting the right-hand side of (6.72) provides us with a linear, positively sloped, curve, on which $dk/dt = 0$ for $k > 0$; it is shown as the $dk/dt = 0$ curve of the northeast quadrant. Above it, (6.72) shows that $dk/dt < 0$, so that (6.70) and (6.71) show that $dr_1/dt < 0$ and $dr_2/dt < 0$, explaining the direction of the vertical and horizontal arrows; exactly similar reasoning explains the direction of the arrows below this curve. Assuming the economy to be in short-run equilibrium all the time, the economy must always be on curve SS, and the arrows on it prove the stability of the long-run equilibrium of the system at E. The level of k at this equilibrium

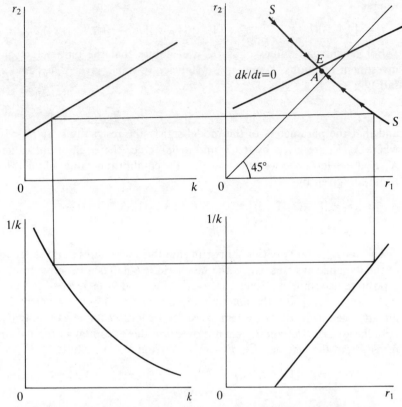

Figure 6.5   Dynamics in the two-sector Kalecki-Steindl model

can be read off from the southwest quadrant and it is then a simple matter to solve for the equilibrium values of the remaining variables.

Four comments on this model are in order. First, given the parameters of the model (that is, the technological parameters, saving rate of capitalists, markup rates, and parameters of the investment functions), the economy will converge to a long-run equilibrium at which it will attain a steady state. Secondly, given arbitrary values of the parameters of the model, there is no reason why in this equilibrium the profit rates of different sectors have to be equal. There need thus be no tendency towards the equalization of rates of profit. It is even possible that over time the rates of profit may diverge as the economy converges to steady state. (An example is shown in Figure 6.5, with the economy starting from A at which $r_2 > r_1$, and the gap between the two widens as the economy converges to E.)

Thirdly, if the parameters of the model obey some specific conditions, the rates of profit may be equal at long-run equilibrium. This will be the case, for example, if the $dk/dt = 0$ line coincides with the 45° line of Figure 6.5. For this to be true we need both $\alpha_1 = \alpha_2$ and $\beta_1 + \tau_1(a_{02}/a_{01})(1+z_2)/z_1 = \beta_2 + \tau_2(1+z_2)/z_2$, which will be satisfied if (1) the investment functions in the two sectors have identical parameters and (2) $a_{02}z_2 = a_{01}z_1$. It is not clear why these conditions will necessarily be satisfied. Of course, even for this model, these conditions are not necessary for rate of profit equalization in long-run equilibrium. We simply require that the $dk/dt = 0$ and SS curves intersect on the 45° line. It would appear, however, that the configuration of the parameters required for this would exist only accidentally.

Finally, it can be argued that the non-equalization of rates of profit in our model results from its specific assumptions. First, it can be claimed that the model abstracts from financial markets and their incorporation will imply profit rate equalization. This is by no means obvious: the incorporation of financial capital in the model can result in the equalization of financial rates of return without equalizing rates of profit. Secondly, it can be said that our result hinges crucially on the specific investment functions used. This is true in the sense that it is possible to write down other investment functions which would force rate of profit equalization in the long run. But it is by no means clear why investment functions should take such specific forms.[30] Thirdly, it can be argued that the non-equalization result follows from the unwarranted fixing of some of the parameters of the model, such as the labour-output ratios and markups. Endogenization of these parameters would require an analysis of technical change,[31] changes in industrial concentration, and changes in the balance of power between different classes,[32] which is far beyond our scope here. If one insists on rate of profit equalization, then our model suggests that the endogenization of these elements into the model would be a promising line to take, although the result of such enquiries need not produce the result being sought.

On the basis of these observations we are led to believe that in the presence of monopoly power, steady state growth need not necessarily be characterized by the equalization of profit rates between sectors.

6.2.4 The Feldman-Mahalanobis-Domar model again

Section 6.1.7 examined a version of the FMD model in which rates of profit rate equalized between sectors, and changes in the investment-allocation ratio by planners required adjustments in the real wage. Here we examine an FMD-neo-Marxian economy in which V is fixed at subsistence, and both sectors fully utilize their capacity, and planners try to fix the allocation parameter Θ even in the short run.

As in the previous FMD model we have (6.29) and (6.30). With k given in the short run, these determine both g^1 and g^2. This economy must also satisfy (6.38) and (6.39), which together imply equation (6.48). Further, the economy must satisfy (6.49). These last two equations solve for the rates of profit in the short run, given by (6.50) and (6.51).

We can examine the movement of the economy over time using

$$\hat{k} = (1 - \Theta)(1/a_{22})(1/k) - \Theta/a_{22}. \tag{6.73}$$

They can also be examined from Figure 6.6,[33] where the g^1 and g^2 curves depict (6.39) and (6.38). The long-run equilibrium of the economy is attained at k^*, where $g^1 = g^2$; it is obvious from (6.73) and the figure that the convergence to this long-run equilibrium is stable. The value of k^* is given by (6.31).

As in the previous FMD model, a higher Θ implies a lower k^*. But in this model, where V is fixed, this implies, from (6.50) and (6.51), a higher r_1 and a lower r_2. If capitalist investment is given by equation (6.42), the planners must find some way of preventing capital flows from the capital goods sector to the more profitable consumption goods sector. If the planners are successful in doing this, they can choose any Θ they desire, and hence any long-run rate of growth which preserves positive and meaningful solutions of the relevant variables. The planners can achieve this by tax-subsidy policies, or by direct state ownership of certain sectors. If they own firms in the investment goods sector, they will have to run them at lower rates of profit than those earned in the consumption goods sector. Thus, rather than reflect government ineffic- iency, this lower profitability will signify that the authorities have chosen to have a higher rate of growth without a reduction in the real wage. In any case, note that in the long-run equilibrium, the rates of profit between the sectors are not equalized, and this is a result of deliberate government policy.

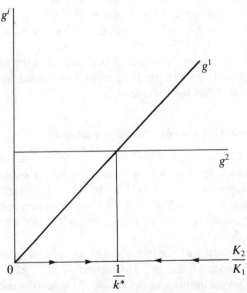

Figure 6.6   Long-run equilibrium in the Feldman-Mahalanobis-Domar model

6.3 Conclusion

This chapter has examined a series of two-sector models in which one sector is the consumption goods sector and the other the investment goods sector. We first considered a concept of long-run equilibrium in which the sectors grow at the same rate with equal profit rates. Using it, we examined long-run equilibrium versions of our four basic models, together with two other that have received some attention in the literature. Following this, we examined some models to explore what happens if we forsake our particular concept of long-run equilibrium. This analysis showed that it is possible for the economy to converge to a long-run equilibrium satisfying our definition, from arbitrary initial conditions. But we also found that it is possible to develop models with monopoly power or state intervention, showing convergence to a long-run equilibrium position with equal sectoral growth rates, but with long-run divergence of sectoral rates of profit.

CHAPTER 7

Some doctrinal issues concerning two-sector models

7.0 Introduction

This chapter uses the long-run models of chapter 6 to examine some issues (that could not be adequately examined within a one-sector setting) which have recently generated much discussion and debate. Section 7.1 considers the debate on the relationship of Sraffian price theory to the neoclassical approach. Section 7.2 examines the nature of the classical theory of prices and shows that it employs a dichotomy which may be inappropriate. Section 7.3 addresses the Marxist debates on whether competition or monopoly is the relevant assumption for analysing capitalist economies.

In examining these issues we will return to the notion of long-run equilibrium (with rate of profit equalization) used in section 6.1. Although this is not always an appropriate assumption (especially for the Kalecki-Steindl model), we do this to make some initial concessions to supporters of the notion of classical competition, and still show that their analysis may be problematic.

7.1 The Sraffa system and the neoclassical model

Sraffa (1960) analysed the production of commodities by means of commodities as a 'prelude to a critique' of neoclassical economic theory. He developed a model in which, given some exogenous elements, it is possible to solve for the prices of production consistent with the classical notion of competition implying intersectorally equalized rates of profit. This analysis has been subject to an enormous amount of scrutiny and been made the basis of attempts – by his followers – to show that neoclassical economics is logically faulty. Neoclassical theorists – most notably Hahn (1975, 1982) – have responded by attempting to show that these criticisms are misplaced, and have gone on to argue that everything that Sraffa analysed can be discussed in terms of neoclassical general equilibrium theory and indeed, that Sraffa's model is a special case of an intertemporal neoclassical general equilibrium model.

Sraffa (1960) assumes given levels of output and technical relationships between inputs and outputs. He also assumes that distribution is given, so that either the real wage or the rate of profit are given but not both (he prefers the assumption of given rate of profit, determined by the interest rate, which depends on monetary factors). Given these, he solves for the prices of production which equalize the rate of profit between sectors. The solution can be found from

$$\mathbf{P}' = \mathbf{P}'\mathbf{A}(1+r) + W\mathbf{a}_0',$$ (7.1)

where $\mathbf{P}$ is the column vector of n prices, $\mathbf{A}$ is the input-output matrix with elements a_{ij} denoting the amount of the ith commodity used for the production of a unit of the jth commodity, $\mathbf{a}_0$ is the column vector of elements a_{0j} which denotes the amount of the only primary factor, homogeneous labour, used for the production of a unit of the jth commodity.[1] We here assume that wages are not advanced prior to production, so that profit does not have to be imputed on the wage cost. It is also assumed that all capital is circulating capital (so that there is no durable capital), and there is no joint production. There are n equations in (7.1), but $n+1$ unknowns: the n prices and either W or r. We may choose any price as the numeraire (Sraffa uses a different normalizing rule), and the n equations can now solve for the $n-1$ relative prices and one distributional variable. Note that Sraffa does not make any assumption regarding returns to scale (that is, how the a_{ij} change with output in sector j) since he takes the output levels of all commodities as given. He then uses his system of equations to derive a downward-sloping relationship between the wage and the rate of profit.[2] He also extends his analysis to allow for the existence of more than one technique (with the economy choosing the technique with the highest rate of profit, given the real wage), for joint production which could also allow for the treatment of durable capital goods, and for land as a second primary factor.

Sraffa's followers have tried to use this analysis to show that neoclassical theory is logically faulty. Thus they argue that neoclassical distribution theory, which uses the concepts of aggregate capital and its marginal product, is invalid for a world in which capital actually consists of different types of produced inputs. Sraffa's analysis shows that the relationship between the return to capital (the rate of profit) and an aggregative value measure of capital (valued at the prices of production) need not be inverse, and this contradicted the neoclassical downward-sloping marginal productivity curve of capital. Marginal productivity theory thus cannot explain the distribution of income, which therefore had to be explained some other way. The Sraffian approach explains it in terms of class conflict, and Sraffa's wage-profit frontier is taken to be a formal representation of this conflict.

Hahn (1975, 1982) strongly criticizes these views. He (see Hahn, 1982) considers an intertemporal neoclassical model with heterogeneous capital goods which are produced inputs, with labour the only primary input, and with smooth neoclassical production functions exhibiting constant returns to scale. Firms and households (which are endowed with labour and fixed amounts of produced inputs at the beginning of the period) are price takers and optimizers, and all markets exist – including futures markets – and clear through price changes. Then, with given initial endowments of labour and the produced means of production, the general equilibrium system solves for all variables (given existence conditions). At this equilibrium, the own rates of return for each produced good are not equal unless the endowments are on the balanced-growth path. If they are, however, the equilibrium solution is the same as that of Sraffa's equations if the input-output ratios used there are those that are solutions for the neoclassical model. Hahn thus concludes that Sraffa's model is a special case of the neoclassical model which in addition assumes (1) fixed coefficients of production (Sraffa of course does allow for alternative techniques in his book, so that this is not quite fair) and (2) the initial endowments are on the balanced-growth path. Therefore, Sraffa's analysis cannot show that the neoclassical model is logically faulty.[3] Moreover, Sraffa's model can be considered to be incomplete, since it can neither determine quantities of output and distribution, nor explain the movement of the economy out of equilibrium with equalized rates of profit; the neoclassical model can do all this.

Even though Hahn's formal analysis is correct, this does not imply that Sraffa's work is of no value, which seems to be Hahn's unstated message. Even if we agree that Sraffa's model is a special case of the neoclassical model, it does reveal the problems associated with the simple-minded neoclassical models with homogeneous capital (which Joan Robinson had long been criticizing), and may have had a part in the development of intertemporal neoclassical general equilibrium models with heterogeneous capital goods.[4] However, it is also incorrect to dismiss Sraffa's work as a special case of the neoclassical model, for several reasons.

First, the neoclassical and Sraffian methods represent very different types of theorizing.[5] The former may be said to be in the Cartesian-Euclidian tradition, and the latter in the Babylonian. The former believes that all theorizing – in this case economic theorizing – can be done using one unified framework; neoclassical general equilibrium theory thus tries to explain the determination of the levels of output, prices, and distribution in terms of one model. The latter believes that economics pertains to a vast area of knowledge which, given the bounded rationality of the analyst and the present state of theory, cannot be analysed in terms of one model. One must look at some parts carefully, treating the other parts as data. Models are by their very nature special cases, rather than being relevant for the whole universe of discourse. Sraffa, following the classical economists, was adhering to this tradition, and he took the output levels and income distribution to be data for his analysis of prices. Another difference in the nature of theorizing, related to the one just discussed, is that the neoclassical method theorized on the basis of counterfactuals (demand and supply functions, for example, are not observables, but consider what might have happened had prices been different), while Sraffa theorizes on the basis of observables, such as levels of output and the distribution of income.[6] Hahn's treatment glosses over these important methodological differences, to which we will return in the next section, and assesses Sraffa from a neoclassical methodological perspective.

Secondly, Sraffa can be thought of as following the classical tradition of discussing only long-run equilibrium positions (perhaps based on the belief that short-run issues are transitory in nature and dependent on uncertainty), while neoclassical theory concerns itself with temporary equilibrium as well as its movement over time.[7] Hahn takes the neoclassical view for granted and criticizes Sraffa for not examining short-run equilibrium positions.

Thirdly, let us agree to be Cartesian-Euclidian in attempting to examine one model which examines the determination of prices, growth, and distribution (which is consistent with neoclassical method) and decide to restrict ourselves to long-run equilibrium positions (consistent with Sraffian, and not inconsistent with neoclassical methodology). We can then use the models of section 6.1 to show that the neoclassical model is a special case of the Sraffian framework.[8] It is in general possible for each one to be a special case of the other because particular models are always representations of more general 'models' which are too complex to write formally, so that it is possible to show that a particular form of one model is a special case of a particular form of the other, and vice versa. Showing that one model may be a special case of another may be useful for purposes of comparison, but should not be made the basis of belittling the model which appears to be a special case of the other.

The part of our general framework given by (6.9) and (6.10) may be thought of as a simple representation of a Sraffian price system. These equations portray an economy somewhat different from that of (7.1), since they consider only two commodities rather than n, and only infinitely durable fixed capital rather than circulating capital completely used up during the period of production.[9] These differences, however, are surface differences that can be removed without affecting our conclusions. The similarity lies in the fact that these equations can solve for relative prices and one distributional variable given the others, as was done by Sraffa in (7.1). Our equations, given (K_i/X_i)'s, r (or V) and a_{0i}'s, can determine the values of p and V (or r). However, they can not only be imbedded in a neoclassical model which determines all the variables under the assumption that all markets (including labour markets) clear with price changes, but also in neo-Marxian, neo-Keynesian and Kalecki-Steindl models. In this sense the neoclassical model can be thought of as a special case of the Sraffian system. While the solution of the neoclassical system is consistent with the values obtained in the Sraffian equations, so are the solutions of the other three systems, and they cannot be said to be generated by the same economic forces, and certainly do not have the same theoretical implications. The Sraffa system can thus be thought of as paving the way for the emergence of theories of growth, distribution, and prices, alternative to neoclassical theory. This is not the same as arguing, of course, that the neoclassical theory is logically faulty.

7.2 The classical dichotomy: the theory of prices with given output and distribution

The classical economists – including Ricardo and Marx – employ a dicthotomy (as defined in section 1.2) which takes the level and composition of output, technology, and distribution, as data and examines the determination of prices in an economy in which the forces of competition prevail.[10] As seen in the previous section, Sraffa (1960) follows them by examining the determination of prices which equalize profit rates between sectors (due to competition) in a system in which input-output ratios (technology and outputs) and one of either the wage or profit rate (distribution) are given. While both the classical theory of prices and Sraffa's work have been much criticized, very little of this criticism directly questions the validity of the dichotomy between prices, on the one hand, and output, technology and distribution on the other. Here we critically examine the nature and general validity of this dichotomy, by first examining the nature of dichotomies in general and then by examining the classical one in terms of the models of section 6.1.

7.2.1 Dichotomies as analytical procedures

Economic theorizing necessitates simplifications. In considering a particular economic problem, one can examine the interaction between only a subset of a large number of factors which can possibly have a bearing on that problem; other factors have to be either left out altogether, or taken as givens in the particular analysis. In section 2.1 we referred to this procedure as the use of analytical dichotomies.[11]

Economists explicitly admit to using dichotomies. For example, Marshall (1890, p. 304) writes:[12]

> [D]ifficulties in economic investigations ... make it necessary for man with his limited powers to go step by step; breaking up a complex question, and studying one bit at a time, and at last combining his partial solutions into a more or less complete solution of the whole riddle. In breaking it up, he segregates those disturbing causes, whose wanderings happen to be inconvenient, for the time in a pound called *Caeteris Paribus*. The study of some group of tendencies is isolated by the assumption *other things being equal*: the existence of other tendencies is not denied, but their disturbing effect is neglected for a time.

This suggests that dichotomies are accepted as a heuristic device, as a step to a more complete understanding of the whole problem being studied. This can be taken to imply that the nature of the dichotomy used does not matter, since it is only a step. We argue, however, that this need not be the case.[13] First, although the factors initially ignored (either being left out or being taken as given) may eventually be introduced into the analysis, this is done usually in a *less formal* manner than when the factors included are analysed. This implies that results derived from this analysis may be quite different from those that could have been derived from a formal analysis which included the initially ignored factors.[14] Secondly, and as mentioned in section 1.2, even if ignored factors are included later, this inclusion occurs at a stage at which the theory without them is already well developed, so that the extensions – consciously or unconsciously – will usually maintain the structure and implications of that theory in tact. Thirdly, if we take the reasonable view that *all* factors can never be introduced, we cannot rest on the hope that eventually we will be able to arrive at the absolute 'truth'.

A brief enumeration of some well-known dichotomies in economics will convince us as to their widespread use as well as to the fact that many theories have been criticized – in the final analysis – for the kind of dichotomy they adopt. According to Keynes, 'classical' theory examines production, consumption, and exchange, as if the economy is a barter one, and monetary considerations are left outside its formal theoretical structure; they are appended later on, sometimes as a version of the crude quantity theory. Keynes's critique of this theory can be viewed as an objection to this dichotomy between real and monetary factors, which makes such theorists underplay the role of uncertainty and overlook the possibility of unemployment without inherent mechanisms for its removal. One interpretation of Keynesian economics is a change in the type of dichotomy used: with prices and wages fixed, the *IS-LM* apparatus explores the determination of output and the rate of interest; price and wage changes are then introduced into the analysis at a less formal level.[15] Most of neoclassical theory assumes that psychological, sociological, political, and technological factors can be taken as data. Thus tastes, social norms, the nature of government policies, and technology are typically taken as given, and interrelationships between narrowly defined 'economic' variables are analysed. The well-known criticisms of neoclassical theory, especially in the study of growth and development, imply a rejection of this dichotomy between 'economic' and 'non-economic' factors.

Given the existence of dichotomies which have been open to a great deal of criticism, and the necessity of using some dichotomies in theoretical work, one may ask if there are criteria by which to judge the validity of a dichotomy. To find such criteria, it is necessary to distinguish between two types of dichotomies used in economics.

One type, which we can call a strong dichotomy, is used when a part of the entire (possible) system is analysed in which factors from outside it are completely ignored. An example of this type of dichotomy is the 'classical' one described above, in which the real part of the system was analysed, money being left out of it altogether.[16] Formally, assume that in principle – the entire system can be depicted by a set of equations, which can determine all variables.[17] A strong dichotomy can be said to be valid when a subset of these equations can be written in such a way that the values of the variables determined in that subset do not depend on the values of the variables in the rest of the system.[18] In actual applications, of course, one has to produce plausible theoretical arguments or empirical 'verifications' showing this independence.

A second type of dichotomy, which we call a weak dichotomy, involves the consideration of a subset of the entire system in which certain factors which are variables in the entire system can be taken as given. The dichotomy used in fix-price Keynesianism is clearly of this type: the price level is considered fixed in the analysis (in, for example, the *IS-LM* analysis). The dichotomy is not strong: changes is the price level will have effects on the subsystem being analysed. The classical dichotomy with which we are concerned here, is also of this type. Formally, assume again that in principle the entire system can be depicted by a set of equations which can determine all the variables in the system. A weak dichotomy can be said to be valid when those factors treated as data in the analysis of a subset of equations in the entire system of equations are not affected by any of the variables in the subsystem, or by any of the other data in the subsystem. This implies that movements of the data of the subsystem do not alter the form of the equations of the subsystem with which we are concerned (other than that directly caused by the change in the data). In actual applications we can take such dichotomies to be valid if the interrelations between the factors we have mentioned are logically or empirically shown to be absent or extremely weak.

The foregoing analysis implies that to examine the validity of a particular analytical dichotomy – whether strong or weak – we need to embed the theory (involving the subsystem of equations) in a complete theory, and to check whether the conditions of validity discussed above are fulfilled. Since, however, the complete system is not known – and that, indeed, is the rationale for using dichotomies – we may check for validity by embedding the subsystem in a less restrictive subsystem in which at least some of the data of the subsystem under examination are converted into variables, and none of the variables are converted into givens. By this method, of course, we can only check for the validity of the dichotomy in a limited sense, since we endogenize only *some* of the data. We call this type of validity 'general' validity.

Our formal definition of a valid weak dichotomy ignores two important issues. First, while we assume that all functional relationships of the entire system are known, in economics, where relationships between very few variables can be said to be well established, this is hardly the case. Secondly, by assuming that all equations are 'satisfied' we have abstracted from movements of variables over time; when such movements are examined, we may see that some variables change more rapidly than others. These issues suggest two additional types of valid weak dichotomies.

First, if the interrelation between them and other elements is either uncertain, or not well understood, or both, these elements can be taken to be exogenous in the examination of a part of the entire system, within which the relationships are better understood. In this case we may take the dichotomy to be valid, and call it 'incognizance' validity. Neoclassical economists often use their (supposed) ignorance regarding the determinants of non-economic factors for treating them as exogenous.[19] Similar arguments justify taking expectations to be exogenous in Keynesian models. The acceptability of a particular dichotomy in terms of this criterion can obviously change with the development of adequate theories regarding factors which previously were not well understood.

Secondly, if some elements change over time more slowly than others, these elements can be held constant for the analysis of the part in which the elements which move more quickly are explained. Movements in the system can then be studied separately when these relatively slow moving elements change. The short-run-long-run distinction used, for instance, by Marshall (1890), is made along these lines.[20] The validity – which we call 'temporal' validity – of a particular dichotomy using this criterion depends on the time horizon for which the analysis is supposed to be relevant, and also on the conditions prevailing in the economy. Thus wages and capital stock may be taken to be exogenous in the short period, but perhaps not so in the long period. The stock of machines may be taken to be exogenous in a model in which employment of labour is allowed to vary for one kind of an economy; for economies with long-term labour contracts the opposite may be valid.

7.2.2 The classical dichotomy

As we have seen above, the classical economists used a dichotomy by which they examined the determination of prices given output, techniques, and distribution, and Sraffa has continued in this tradition. As Roncaglia (1978) writes:

> Sraffa has chosen the relationship between production prices and distributive variables (rate of profits and wage rate) as the objects of the analysis. All other variables (technology, levels of output, firm structure of all industries, etc.) are taken as data of the problem ... (t)his choice does not imply an *a priori* refusal of the possibility of analyzing the problems of technological development, levels of output, strategy of firms, etc. This choice stems from the necessity of analyzing the different problems one by one, and each in isolation. The necessary assumptions and methods of analysis are not necessarily identical for all problems; for each of them only what is relevant should be included, leaving aside those elements which as Ricardo said, simply 'modify' the analysis but do not change it substantially.

As discussed above, with these assumptions Sraffa showed the existence of prices of production, and also the existence of a downward-sloping wage-profit frontier.

Several neo-Ricardians explain why, in analysing prices, the other elements can be taken as data. Garegnani (1984) writes:[21]

> The multiplicity of these influences and their variability according to these circumstances was in fact understood to make it impossible to reconduct them to *necessary* quantitative relations like those, studied in the core ...

> [F]or example important changes in the real wage may have a multiplicity of effects on aggregate demand, the intensity of which will depend on the particular circumstances in which they occur and cannot, in our present state of knowledge, be reconducted to known functional relations of sufficient generality and persistence. If this is admitted, it will appear that a general determination of outputs simultaneously with relative prices is impossible, and the basic procedure can only be that followed by the classical authors. They analyzed changes in prices and outputs by what we may call two distinct logical stages. In the first, the effect of the change in the real wage was examined while taking the outputs as given. In the second stage, the possible effect on *outputs* of the initial change was analyzed in accordance with the circumstances of the case under consideration, jointly with its possible secondary effects on prices and distribution, in the case of non constant returns to scale. [Emphasis in original].

The argument is that while the relationships within the core are well known and definitive, the same is not true for the other relationships, which are diverse and not subject to general, well-understood, laws. The validity is thus defended on the lack of knowledge.

There has been much criticism of the classical method as formalized in Sraffa's work. We briefly examine two of these, and show that they are related to the question of validity of the dichotomy.

First, while Sraffa (1960) points out in his preface that he does not need to assume constant returns to scale, a group of critics insist that, to make his analysis meaningful, he needs the assumption. Without it, many of Sraffa's central propositions do not hold.[22] This is because if output levels change, with non-constant returns to scale, the input-output ratios also change. Thus Sraffian equations with given input-output ratios cannot be taken to depict equilibrium states unless they are stationary states. Even with stationary states, the equations cannot be used to examine the impact of distributional changes because they change the composition of demand, the levels of output and hence input-output ratios.[23] Those who deny the necessity of the assumption argue that Sraffa's method is not to study actual long-period positions, but to examine an economy at a point in time at which actual prices diverge unsystematically and insignificantly from prices which equalize the rates of profit (corresponding to the dominant techniques of production), which are the centres of gravitation for actual prices about which nothing definite can be said.[24] About the effects of changes in wages, Sraffa's wage-profit frontier can be taken to give only first round effects, without considering later rounds in which output changes; it therefore just studies notional, not actual, changes. However, the usefulness of studying such snapshot pictures for an economy which always undergoes changes, and of examining the notional changes, can legitimately be questioned, and begs the question of the validity of the classical dichotomy.

Secondly, there is the criticism that the Sraffa system does not determine all the variables of interest in the classical economy, since it takes output levels, income distribution, and technology to be given. As discussed above, neoclassical economists claim the superiority of the neoclassical general equilibrium model over the Sraffian framework on this ground, because the former can solve for outputs and distribution and the latter fails to do so. Neo-Ricardians, on the other hand, claim that the Sraffa system does not *fail* to solve for outputs and distribution, it does not *intend* to do so for the reasons discussed above. They claim that this makes the classical framework less restrictive than the neoclassical supply and demand framework. As Bharadwaj (1978) puts it:

> it does not commit itself through its theoretical structure to any form and direction of change ... the classical theory is not constrained to permit only some specific changes of the many possible ones as alone consistent with theory. Thus it does not have to presume more than is necessary for the limited objective of determining relative values at one 'observed' position of the economic system.

Some neo-Ricardians extend the Sraffian framework to analyse the determination of the levels of output. They seek to demonstrate that, whatever the levels of output, the Sraffian system can solve for prices, and the Keynesian adjustment of output which brings saving and investment to equality can provide the theory of quantities.[25] The approach (Garegnani, 1976, 1978–9, Milgate, 1982, Eatwell and Milgate, 1983) follows Sraffa in confining attention to the long period, but it is not always clear whether the long-period equilibrium *levels* of quantities (which suggests a notion of long period stationary state) or *growth rates* are determined. If valid, this approach vindicates the classical dichotomy and also shows that the neoclassical criticism is unfounded: the Sraffa system extended to included growth rates and distribution as variables can be 'closed' in ways other than the neoclassical one.

7.2.3 *The validity of the classical dichotomy*

As mentioned above, if we take the a_{0i}'s (technology), one of r and V (distribution), and X_i/K_i (output levels), as given, (6.9) and (6.10) determine p and V or r.[26] However, the question arises as to whether the use of such a dichotomy is valid.

If we suppose that technology is given, and if the true model of the economy is either neoclassical, neo-Marxian, or neo-Keynesian, and our other assumptions are satisfied, the procedure seems valid. If we are interested in the determination of the relative price, we can examine only (6.9) and (6.10). All these models assume (6.13) and (6.14) and determine one of r (neoclassical and neo-Keynesian cases) or V (neo-Marxian case) elsewhere in the model and use these equations to solve for the relative price and the remaining distributional variable. Changes in any of the parameters will change the equilibrium values of the variables, but we are justified in using this dichotomy, provided that we are interested in relative price determination and not in the determination of growth rates or distribution. Distribution can be taken as a true parameter in the prices-of-production subsystem. If n increases in the neoclassical model, for example, g^* and r^* increase. The effects of this change in r^* on p^* and V^* can be studied by considering the parametric variation in r^* in the prices-of-production submodel. Similarly for changes in class struggle affecting V in the neo-Marxian model, or 'animal spirits' in the neo-Keynesian model. In particular, we have seen that with (6.13) and (6.14), and with technology given, there is an inverse relation between r and V for each of these models obtained from (6.9) and (6.10): this is Sraffa's wage-profit function.

These same conclusions cannot be arrived at if the true model is a Kalecki-Steindl one. In this case (6.13) and (6.14) do not hold. V^* is determind by (6.21), and seen to depend on technology and the markup rate in the consumption good sector. The effects on relative prices (and the profit rate) of a change in V due to a change in the markup rate cannot be examined from (6.9) and (6.10) holding the X_i/K_i constant simply because the latter will change in a predictable manner when the markup changes, as analysed above. If we want to examine the effects of the change in markups on relative price we must look at (6.22), not (6.9) and (6.10). In this model (6.22) – and not the prices of production equations – 'determines' relative prices, since it solves for p^* in terms of parameters. A valid weak dichotomy is thus to look at (6.22).

It should be pointed out that, if the X_i/K_i's are held constant, (6.9) and (6.10) will still imply a downward-sloping wage-profit frontier. However, in this model, a rise in the real wage does not imply a fall in the rate of profit; the wage-profit frontier is actually upward rising! As already shown, a higher real wage in this model is associated with a higher rate of profit and a higher rate of growth.

It is thus clear, that, if the true model is a Kalecki-Steindl one, the classical dichotomy is an invalid one if we insist on a weak dichotomy. But we have argued earlier that weak dichotomy may be too strong a requirement. The question arises as to whether the dichotomy can be defended on the two other grounds discussed above.

If we did not know how the X_i/K_i's were determined, then on grounds of ignorance, we could take them as exogenous. Especially after the development of theories of effective demand by Keynes and Kalecki, such a claim seems unfounded. The classical economists did not examine the pricing and output decisions of firms, but once we examine the ideas developed by Kalecki we find that the classical approach implicitly assumes that firms are price takers (which guarantees full capacity utilization) or that demand is enough to keep the economy at full capacity utilization in equilibrium. If the Kalecki-Steindl theory is valid, and it is known to us, the classical dichotomy cannot be defended on grounds of ignorance about output levels.

Regarding the question of the time horizon, the classical dichotomy could be defended if output levels change slowly relative to prices (although such a defence may not appeal to those neo-Ricardians who have a long period equilibrium theory mind). Such a defence is possible if price variations clear markets and outputs change slowly due to capital accumulation. In the Kalecki-Steindl framework, however, quantities vary to clear markets, while prices are fixed by the stable degree of monopoly power, so that this defence is invalid. The time horizon could be used as a defence in another way by arguing that the classical theory dealt with long-run equilibrium positions in which output was at full capacity. In this interpretation, the Kalecki-Steindl model is invalid in the long run, a position that we have already argued against, in sections 3.4.1 and 6.1.5.

Though Garegnani (1984) suggests that we have better reasons for accepting the classical dichotomy today than in their time, our criteria suggest the contrary. The relative unimportance of large firms which set prices and of fixed capital (which increases the chance of excess capacity) could have ensured the validity of the dichotomy when the classical economists wrote. But in the days of giant corporations and of fixed capital, the dichotomy (although *logically* sound) seems unreasonable. Our Kalecki-Steindl model embodies a different dichotomy (in which technology and the degree of monopoly determine prices) which can easily be defended according to our criteria.

We may address some of the issues mentioned above, relating to the classical dichotomy, in terms of this analysis.

In the debate on the constant returns to scale assumption, we saw that without it, the input–output ratios depend on output levels. In this case, even if the economy operates at full capacity utilization, the property of weak dichotomy which allows the examination of price determination using the prices-of-production framework breaks down. Changes in output will change techniques as well, and changes in distribution would change output as well as techniques.[27] What Sraffa's critics suggest is that he must assume constant returns to scale to achieve a strong dichotomy regarding output levels (changes in them have *no* effect on relative prices), and a weak dichotomy for distribution.

Our analysis has two implications for this debate. First, if the economy operates with full capacity, the classical dichotomy can be defended even with non-constant returns to scale. Our lack of firm knowledge regarding the nature of returns to scale can be used to justify the simplifying assumption of given input – output techniques. Otherwise, slow changes in input – output coefficients (due to slow changes in output due to the fact that scale changes take time, for instance) can be used as a justification for studying short-run first-order effects. Secondly, if the economy does not operate at full capacity utilization, even the assumption of constant returns to scale will not rescue the classical procedure. Our Kalecki-Steindl model, which assumes constant returns to scale and allows excess capacity to exist, has a positively sloped wage-profit locus.

Next consider attempts at supplementing the Sraffian apparatus with a Keynes-type analysis to have a theory of quantities as well as a theory of prices. Given conditions of steady state growth, it seems too special (a stationary state is required) to expect to determine output levels, we shall interpret 'quantities' to be growth rates of output. Our analysis suggests that there are dangers involved in appending a quantity-determining theory to the Sraffian model of price determination.

The method works with models of full capacity utilization, since in them, the prices-of-production submodel solves for relative prices and one distributional variable given the other, and growth rates and the other distributional variable can be determined with a set of additional equations, one of which may be a desired accumulation function and a saving-investment equality as in the neo-Keynesian model. However, the Kalecki-Steindl model shows that, if we introduce a particular quantity and income distribution determining apparatus in a model in which Sraffian pricing equations appear, the pricing equations may be radically altered. The parameters of the price-determining submodel, in particular, the input–output ratios, change in the model with variations in the distribution of income and the utilization of capacity. It is therefore incorrect in general to suppose that prices are determined in a Sraffian manner, and quantities can be determined using some other theory.

7.3 Competition, monopoly power, and the uniform rate of profit

The classical economists – Smith, Ricardo, and Marx – analyse the laws of motion of capitalist economies in terms of their concept of competition. Following changes involving the concentration and centralization of capital, a tradition of Marxist economists, which we call the monopoly power school, argues that the laws of competition have to be replaced by those of monopoly capitalism, according to which the capitalist system is ultimately regulated by the balance of power relations between workers, capitalist firms, and the state. While accepting the empirical facts regarding changes in industrial organization, another group of Marxist writers argues that such tendencies merely intensify the forces of competition in a capitalist economy, so that the framework of the classical economists, with their concept of competition, and one which can be formalized using the prices of production approach, remains the appropriate one for the analysis of modern economies with large firms. This section reviews the notions of competition and monopoly power and the debate between the two divergent Marxist views, and argues – relying mostly on the Kalecki-Steindl model of section 6.1.5 – that the industrial changes which the monopoly power school emphasizes may well imply that the theory of prices in a capitalist system has to be radically altered.

7.3.1 Competition and monopoly power

Though several earlier writers had discussed the concept of competition and analysed its implications,[28] it is the contribution of Adam Smith to make it the centrepiece of a general theory of prices. While Smith takes competition to involve the mobility of labour, capital, and land between different uses, equalizing wages, the rate of profit, and rent over the economy, respectively, Ricardo emphasizes the distinctive role of capital in competition. Marx also takes competition to imply intersectoral mobility of capital, which tends to equalize the rate of profit in different sectors, but he formulates a more general and dynamic concept of competition.

Apart from differences regarding what regulated market prices (natural prices for Smith and 'almost' labour values for Ricardo are replaced by Marx's prices of production, given by average costs of production and the average rate of profit equalized between sectors by competition), and the much broader role it plays in his theory (being intimately connected to the production, realization and distribution of surplus value), two developments should be highlighted. First, competition does not, according to Marx, bring about a smooth process of adjustment and convergence to centres of gravitation, but results in deviations from them due to disequilibrium of demand and supply, barriers to the entry and exit of financial capital, and differences in the efficiency of individual units of capital. This discussion was far more general than the examples given by Smith and Ricardo to show that for some commodities, in rather specific instances, competition may not equalize the rate of profit. Secondly, Marx (see especially Marx, 1857–8) took an historical view of competition, relating it to the rise to dominance of the capitalist mode of production: competition developed with the development of capitalism, removing legal and extra-economic impediments to capital mobility.

To conclude this examination of the classical theory of competition we note the following. First, for Smith, Ricardo, and Marx, competition refers to the mobility of capital between different sectors, and the resulting tendency towards an equalization of the rates of profit. Secondly, competition, for the classical economists and their precursors entails active competition among sellers and individual capitals. Thirdly, although Smith and Marx discuss the role of individual capitalists, the behaviour of individual procedures as sellers and bargainers is hidden under their role as agents of capital mobility.[29] Fourthly, and as a corollary of this, the classical economists do not have a treatment of the effect of industrial structure of a particular industry on the behaviour of individual sellers. To be sure, Smith and the others do point out that restrictions on entry would result in high prices, due to the fact that production would be low; but they do not examine carefully the nature of decision-making regarding prices and quantities by individual firms, and how these decisions depend on the number of competing sellers. By not discussing such issues, their analysis seems to imply that industrial structure is of no consequence for prices. Finally – and this is clear from the prices of production formalization – the classical approach analyses prices without analysing production levels and their rates of growth in the capitalist economy; we have already discussed this issue of the classical dichotomy in section 7.2.

The marginalist theorists also use the notion of competition, which has been inherited by the modern neoclassical economist as the concept of perfect competition. Instead of describing a process towards equalized profits rates, this concept makes several precise assumptions about the structure of markets in an economy, as follows: agents are 'atomistic' in the sense that there are many buyers and sellers, so that each one perceives that they can trade any quantity they want to at the going price, each product is homogeneous, agents have perfect knowledge, and factors and agents are perfectly mobile in the sense that there are no barriers to their mobility from one use (or market) to another. One implication of these assumptions is that each agent is a price-taker.

Several comments may be made to distinguish the neoclassical from the classical notion of competition. First, while the classical notion refers to a process, the neoclassical one refers to a state. Indeed, efforts to extend the latter notion to deal with disequilibrium and dynamics are wrought with considerable difficulty: without market clearing, agents cannot be expected to believe in their atomistic status, and since all agents are price takers the fictitious auctioner must be introduced to change prices.[30] Secondly, the classical notion does not involve the assumption of atomistic or many agents, and allows firms to change prices, and to compete with each other actively; the neoclassical notion rules out any kind of active competition involving business rivalry. Thirdly, the neoclassical notion is much more precise, and therefore requires the assumptions of perfect knowledge and homogeneous product; the classics did not make any analogous assumptions explicitly, but to the extent they make one commodity have one price, they do require them. Fourthly, the neoclassical notion, unlike the classical one, carefully spells out the behaviour of a firm (and other agents): it maximizes profits as a price-taker, and can enter and exit in response to profits. Finally, the perfect mobility assumption of neoclassical perfect competition comes closest to the classical idea, but there are differences. For instance, in the classics, the mobility emphasized is that of finance capital, and it does not necessarily imply the movement of firms or factors of production (especially 'capital') in a neoclassical sense.[31]

Important changes took place in advanced capitalist economies towards the last decade of the nineteenth century and the first decade and a half of the twentieth century. Industry, which was earlier composed of numerous small family enterprises, increasingly came into the hands of giant corporations.

Marx himself shows an early understanding of these developments when he writes that accumulation goes hand in hand with the twin processes of concentration (involving the absolute growth of individual capitals which accompanies the extension of social capital as a whole) and centralization of capital (involving rapid relative growth of particular firms which occurs with the absorption and liquidation of a large number of smaller firms). This fusion of capitals is partly brought about through economies of scale which allow larger capitals to have cost advantages over smaller ones, and is speeded up by the credit structure of capitalist society. After Marx's death, Engels also described the genesis of trusts and corporations in European countries, as well as the importance of the industrial securities market for these developments. Other writers, including Hilferding (1910), Bukharin (1915), and Lenin (1917) within the Marxist fold and Veblen (1904, 1920, 1923) outside it (although influenced by Marx) also described and analysed the rise of monopoly capital. A consensus emerged among these writers that, due to these developments, a proper analysis of the nature of capitalism has to proceed in terms of power relations than in terms of the forces of competition. However, these early Marxist analysts did not develop a theoretical apparatus for undertaking this analysis. Theorists in the neoclassical tradition, such as Robinson (1933) and Chamberlain (1933), also responded to the changes in the nature of capitalism, and developed theories of monopoly and monopolistic competition, although in a partial equilibrium, microeconomic framework.

Theoretical developments from the Marxist perspective subsequently occurred in the work of Kalecki (1939, 1943, 1954, 1971). He argued that, with the rise of large firms and the demise of competition, firms fix prices and respond to changes in aggregate demand by changing their levels of output (as discussed in chapters 2 and 3). In discussing aggregate demand Kalecki gives a prominent role to investment spending, and with savings (undertaken mainly by capitalists) adjusting to it, it determines the level of output of the economy. According to Steindl (1952), who continued in the same direction, the utilization of productive capacity is fundamental to the investment planning of monopolistic firms; confronted with a decline in demand, firms cut back output (as in Kalecki) rather than prices, but the resulting fall in output will reduce capacity utilization, and therefore reduce investment and growth rates. The resulting secular stagnation was directly seen to be related to monopoly capital; a simple one-sector formalization of these ideas was examined in chapter 2. The same theme of monopoly power resulting in the breakdown of competition, regulation of the dynamics of the economy by power relations, and the tendency to stagnation, is found in the writings of other Marxist writers such as Baran, Sweezy, Dobb, and Sherman, and also in the work of Sylos-Labini, following the Keynesian ideas of Hansen.[32] Baran (1957) and Baran and Sweezy (1966), for example, examined stagnation in advanced countries due to the rise of monopoly power which reduced investment rates by slowing down the rate of technological change and lowering output. While the analysis has some problems (Dutt, 1988a), it can be formalized with the Kalecki-Steindl model.

We conclude this review of the monopoly power view with four comments. First, several theorists felt that since capitalist economies had changed in structure the laws of competition had to be replaced by the laws of regulation by monopoly power in theoretical analysis. Secondly, they gave much attention to how increasing monopoly power affected the behaviour of large firms, departing from the classical practice of paying little attention to the firm, and coming closer to the neoclassical practice of explicitly examining firm behaviour. Thirdly, the analysis of pricing behaviour, at least in the work of Marxist economists, was closely tied up to the determination of output and rates of growth: monopoly power was related to lack of demand and stagnation. Finally, the analysis of these theorists was usually macroeconomic in nature, so that no alternative to the prices of production framework, which showed how relative prices were formed in an economy in which competition reigned, was offered. To the extent that one of the aspects of monopoly power was barriers to the mobility of capital, some tended to argue that a uniform rate of profit was no longer to be expected. Thus Sweezy (1942) wrote that 'the transition from competition to monopoly brings with it an increase in profits and hierarchy of profit rates ... [instead] ... of the tendency to an equality of profit rates which is a characteristic feature of competitive capitalism'.[33]

7.3.2 Classical competition versus monopoly power

The monopoly power school has been challenged by other Marxist economists who argue that, despite the emergence of monopoly power, capitalism is regulated as before by the self-expansion and competition of capital.

This group finds support in the work of Lenin who, in spite of describing the process of concentration and genesis of monopoly power, in the words of Semmler (1984) argued that 'competition is *not* abolished by concentration and collusion, but *resumed at a higher level* ... oligopolization as well as increasing competition and rivalry in the production, circulation and banking sectors ... [were] ... a necessary tendency in capitalist development'. Lenin's ideas are echoed in Varga's (1948) conclusion (from his study of the monopolistic stage of capitalism in the 1930s) that concentration, barriers to capital mobility, and collusion did not abolish competition but led to the creation of oligopolistic groups and to competition and rivalry among them. According to him, monopoly prices and monopoly profit rates were not overall phenomena, but special cases. Following Lenin and Varga, Shaikh (1978, 1980, 1982), Weeks (1981), Clifton (1977, 1983), and Semmler (1982, 1984), among others, have systematically explored this position. The thesis of this school can be summarized in terms of the following four propositions.

1   The monopoly power view results from a misunderstanding of the classical notion of competition. As discussed before, classical competition does not assume the existence of many sellers so that concentration and centraliz-

ation, and consequent changes in industrial structure, do not negate it. It is neoclassical perfect competition, which assumes the existence of many sellers, which became obsolete.[34]

2 The institutional changes stressed by the monopoly power theorists have actually heightened competition in the classical sense. This finds direct support in Marx's views on the historical development of competition, mentioned above. Mobility of capital in the classical sense involves the intersectoral mobility of financial capital and not necessarily the movement of *firms*, so that arguments relating the size of firms to restrictions on the mobility of capital are irrelevant. In fact Clifton (1977) argues that the rise of giant firms facilitates the flow of capital, since they are able to move capital in search of profits more easily, and have better access to information on profit-making opportunities. Further, the growth of communications networks *pari passu* with the development of capitalism facilitates the transfer of capital and spread of information. Clifton (1983) also argues that the rise of the institution of the multi-divisional corporation, controlled financially by the general corporate office, intensifies the forces equalizing the rate of profit: often divisions exist in different sectors, and the corporate office makes investment decisions by examining the rates of profit in different divisions. Finally, Auerbach and Skott (1988) argue that large firms overcome barriers due to geographical distance – both in locating their production operation and finding markets – more easily, and this also tends to equalize profit rates.

3 Empirical arguments in support of the monopoly view showing the preponderance of price-setting practices and markup pricing are erroneous. Clifton (1983) argues that the spread of price fixing strategies among large firms does not imply monopolistic price setting, but the greater ability of firms to set prices by examining their costs of production, signifying that prices are more quickly adjusted to the prices of production. Semmler (1984) shows that the prices of production equations can be manipulated to produce markup pricing equations of the type estimated by the monopoly power view, thereby arguing that differences in markup rates are not contradictory with the equalization of rates of profit, reflecting only differences in capital-output ratios between industries.

4 The overwhelming evidence regarding rates of profit differentials between sectors (Sherman, 1968, Pulling, 1978) are incorrect or irrelevant for the debate. Semmler (1984) argues that the differentials are not incompatible with the classical notion of competition which treats the prices of production as centres of gravity. As mentioned above, Marx analysed why rates of profit could diverge both within and between industries. The empirical results showing differentials also do not show a stable and persistent hierarchy of profit rates related to industrial concentration; entry barriers seem to be a necessary condition for them to occur. These and other findings Semmler finds to be quite consistent with Marx's

analysis. Glick (1985) argues that the data that monopoly theorists cite imply only short-run differentials in profit rates; proposing an alternative test of the existence of long-run differentials, he shows that, while the passage of time does not completely eliminate the differentials, their size and persistence has been overstated by the short-run studies.

The implication of all this is that this group – which we will call the competition school – is quite content to use the prices of production framework, together with its implied concept of competition, for the analysis of prices in capitalist economies.

7.3.3 *Implications of the Kalecki-Steindl model for the debate*

We here argue against the position of the competition school using the two-sector Kalecki-Steindl model (referred to as the model) of section 6.1.5. We do not argue that the competition school is necessarily wrong, but only that it is possible to have economies with large firms in which the prices of production framework is inadequate for understanding the formation of prices, and that an alternative framework focusing on the balance of power relations – as suggested by the monopoly power school – is required. Our argument proceeds in terms of the following remarks.

1   The model, with classical competition, but with firms setting prices and adjusting quantities in response to changes in demand, shows that there is no necessary inconsistency between classical competition and monopoly power. The former refers to the process of equalization of the rate of profit through the mobility of finance capital (reflected in (6.9) and (6.10)), while the latter, as interpreted here, refers to the ability of firms to set prices (as reflected in (6.18) and (6.19)).

2   The question arises as to which of the equations – the prices of production or markup pricing equations – 'determine' prices in our model, and thus what forces are relevant for studying the formation of prices. If we mean by 'determination', the solving of values of the variables to be determined in terms of some equations where the only other elements are the parameters of the model, then the prices of production equations (6.9) and (6.10) do not determine prices since they contain also output-capital ratios which are variables in the model, but (6.18) and (6.19), rewritten as (6.22), do since the relative price is seen to depend only on technology and the degrees of monopoly power which are the parameters of the model. And, if these are the equations that determine prices, then the balance of power relations between the workers and capitalists and between the capitalists and the state determine prices (since they affect the markup rates), as argued by the monopoly power school. The school may therefore have been right in suggesting that the theory of prices had to be changed as a result of the rise of monopoly power.

3 If there are many price-taking firms the economy will operate with full capacity. In this case, prices will be determined by (6.9) and (6.10), the prices of production equations (after substitution from (6.13) and (6.14)). Thus the prices of production provides an adequate theory of prices for the early stage of capitalism if it had markets with many sellers.

4 If there are few sellers in each market, but they work at full capacity, the prices of production will still be an adequate framework for studying the determination of prices. This implies that the changes in industrial structure need not out of logical necessity imply the theoretical obsolescence of that framework. Whether or not this happens depends on whether or not firms operate with excess capacity and with fixed prices. Notice that we do not need excess capacity in both industries for the prices of production not to determine prices: its presence in only one could introduce an additional capacity utilization variable in them (as shown in section 6.16).

5 That the emergence of large corporations increases competition in the classical sense may be true, but it does not change the nature of our conclusions, since our model assumes classical competition and yet shows how monopoly power regulates capitalism. Even the argument about large firms overcoming geographical barriers and destroying the market power of others need not logically follow, since they could invade markets with several smaller, vigorously competing firms, and destroy whatever competition existed and they could collude (even implicitly) with other large firms to partition the world into several markets.

6 Attempts to show that there is no stable hierarchy of profit rates and that classical competition does not require equalized profit rates all the time may be valid, but have no bearing on the debate on whether the rise of monopoly power necessitated the abandonment of the prices of production framework. This has been hinted at by Sherman (1983), but many, including Semmler (1982) and Glick (1985), continue to believe that it is central to the debate. Perhaps it is all for the best that this is not the case, since the empirical discussions perhaps convince all but the completely committed that neither side has won; everything depends on how one actually measures one's theoretical categories, what time periods are chosen, and how stringent one is willing to be in the interpretation of 'equalization'.

7 Attempts to make price-fixing practices and Kaleckian markup pricing equations consistent with the prices of production equations are quite possibly flawed. Semmler (1984) is right in arguing that markups can be calculated from prices of production equations and do not necessarily show the existence of monopoly power or unequal rates of profit. He also writes that (Semmler, 1982):

> (t)he mark-up over prime cost – in Kalecki's theory a measure of the degree of monopoly power – might be only another expression for the *uniform* profit rate... This markup must be different in industries where the capital/output

ratio ... is different, whereas the profit rate ... may be the same in all industries.

But he simply assumes that the economy operates at full capacity, and that markups are determined once prices and costs are known from the prices of production equations, as in the version of our model with full capacity utilization. But he misses the point of Kalecki's analysis which shows that, with excess capacity, the markup equations (with exogenously determined markups depending on the balance of power relations) determine prices. The prices of production may be consistent with these prices if rates of profits are actually equalized, but they have to be evaluated at the macroeconomically determined capital-output ratios which are not known to us without knowledge of the rest of the model.[35]

8   The failure of the competition school to appreciate the role of monopoly pricing practices in price determination can be argued to be due to an uncritical return to the method of classical price theory, which, as noted above, takes as data the level and composition of output, technology, and distribution, and also does not consider the individual firm and its decision-making in an adequate manner, except as an agent of capital mobility. The first characteristic, the classical dichotomy between the theory of price and the theory of output and distribution, implies that it is acceptable to use the prices of production equations to determine prices. We have argued in section 7.2 that this is not permissible if excess capacity exists, since, with levels of output responding to changes in demand, it is not useful to have a theory which takes output levels as data. Neoclassical theory broke this dichotomy, and the monopoly power theories did the same. In fact they did so with a vengeance, generally being concerned with rates of growth, investment, output, and income distribution – that is, macroeconomic categories – rather than relative prices. The second characteristic raises the possibility that the functioning of the economy in the classical system may be inconsistent with certain types of firms' behaviour. The central feature of monopoly power, as interpreted here, is the ability of firms to set prices, rather than being passive price-takers. Price-fixing behaviour implies that firms must be able to vary output in response to changes in demand. Without explicit attention to output and price decisions of firms, classical theory is incapable of analysing the implications of such changes. Neoclassical theory considers decision-making units much more carefully, and the theoretical implications of monopoly power (imperfections in competition) are explored in terms of that theory. Monopoly power theorists, especially Kalecki, also carefully analyse the behaviour of firms, but do not incorporate that analysis in a model which examines the determination of relative prices. By attempting to do that in our model we have found that classical theory, by ignoring the decisions of individual producers, makes implicit assumptions: that firms are atomistic price-takers, or they operate with full capacity. Thus

their price theory is inconsistant with price-fixing and quantity-adjusting behaviour of firms with monopoly power.

We may thus conclude that changes in the structure of capitalism involving the concentration and centralization of capital may well have implied that the classical theory of price determination by the forces of competition, by which is meant the examination of the prices of production equations with equalized rates of profit, have to be replaced by a theory which examines the determination of prices by the balance of power relations between capitalists, workers, and the state. It should, of course, be borne in mind that our arguments are made with the use of a simple two-sector Kalecki-Steindl model, since our general framework on which it was constructed is a simple one.

7.4 Conclusion

This chapter has examined several issues connected with classical price theory, particularly as interpreted in Sraffa's analysis, using the steady-state models developed in the previous chapter. It has argued that Sraffa's analysis should not be dismissed as a special case of neoclassical economics, and can be seen as opening the way for the emergence of theories alternative to the neoclassical one. It has examined the nature of the dichotomy used in classical and Sraffian price theory, and has argued that it may be an inappropriate one, given the existence of monopoly power. Finally, it has argued that, as a result of the rise of monopoly power in capitalist economies, the analysis of prices may have to pay attention to the balance of power between classes rather that classical competition as formalized in the prices of production framework.

We conclude with the comment that in this and in the previous chapter we have interpreted long-run equilibrium as positions of steady state in which both consumer and investment goods sectors grow at the same rate. It is well known, however, that during the course of growth different sectors grow at different rates due to a variety of reasons related to production and demand conditions.[36] While we have not examined the dynamics of such phenomena – which can be called unbalanced growth or uneven development – our analysis of steady states can be extended to consider specific mechanisms of uneven development in which the long-run equilibria change over time to cause non-proportional growth. We shall examine how this can be done in the context of international development in later chapters, after analysing steady states in the international economy in the next one.

CHAPTER 8

Alternative models of North-South trade

8.0 Introduction

This chapter extends our closed-economy models to allow for foreign trade, by examining models of trade between two regions, which we call the North and the South. In addition to allowing us to consider the structure of open economies, this enables us to explore the implications of trade for international development. The development of these North-South models follows from our interest in focusing on one of the major issues of international growth and development, that of uneven development.

There is a voluminous literature on issues relating to the trade between rich and poor nations, or what has been called North-South trade. Much of this literature has been concerned with the phenomenon of uneven development, in which the rich North becomes richer and the poor South becomes poorer as a result of their economic interaction, an important form of which is trade. The contributors to this discussion comprise a diverse group, and include Amin (1977), Baran (1957), Braun (1984), Emmanuel (1972), Frank (1975), Galtung (1971), Kaldor (1979), Lewis (1969, 1978), Myrdal (1957), Prebisch (1963), Singer (1950), and Wallerstein (1974).

This discussion is not formal, and has hardly any connection with the formal theory of international trade which is dominated by the neoclassical Heckscher-Ohlin-Samuelson (HOS) theory, which has tended to strengthen the presumption that countries gain from trade. The major differences between the uneven development literature and formal trade theory seem to lie in the nature of assumptions reflecting their world view, and the way they evaluate the consequences of trade. Formal trade theory typically assumes that economies are perfectly competitive, and in particular, that all markets are cleared by price variations.[1] The uneven development literature, not being formal, does not make its assumptions clear; however, the discussions show that it does not share the assumptions of the HOS model.[2] Formal trade theory evaluates the consequences of trade by examining utility streams derived from consumption bundles, while the uneven development literature usually discusses long-run development tendencies focusing on patterns of capital accumulation.

A number of formal models of North-South trade have recently been

156

developed to formalize some of the verbal arguments of the uneven development school, dropping neoclassical assumptions where necessary, and focusing on accumulation patterns rather than utility streams. Among these are the models of Findlay (1980, 1981), Taylor (1981, 1983), Dutt (1984b, 1987c), Vines (1984), and Conway and Darity (1985).[3] All of these models assume a two country world, with a given pattern of complete specialization, but they make different assumptions about the structures of the North and the South. They consequently have different analytical properties and show different responses to parametric changes; they may thus have different implications for the role of particular mechanisms in causing uneven development.

This chapter follows our earlier procedure of providing a general framework for studying different models, so that the analytical properties of each model can be explored and compared, and the dynamics of uneven development better understood.[4]

Four preliminary remarks are in order:

1 Except for a few remarks regarding short-run behaviour, our approach will be to examine long-run (steady state) equilibria, and to study the effects of shifts in it due to parametric changes. This is not to deny the importance of short-run phenomena, which has been stressed earlier. Our rationale for studying long-run equilibria arises from the fact that uneven development issues are typically long-run issues, and our different models are compared easily using this approach, their short-run behaviour possibly being diverse. The short-run behaviour of several models of the type to be explored have been studied elsewhere, and the rest can be explored in the way suggested below if the models are found interesting.[5]

2 We found in chapter 2 that for the closed economy it is not difficult to construct a general framework with little economic content (the only economic assumption there being the type of technology). In our general North-South framework to be used here, on account of its greater complexity, some more special (simplifying) economic assumptions will be used. These include, but are not confined to, the South being unable to produce capital goods, and Southern workers consuming only Southern goods. This specificity of the general framework rules out the examination of several kinds of interesting models, some of which have already been studied, but hopefully is general enough for analysing a variety of models and illustrating the central features of the approach adopted here.

3 Chapter 2 examined four basic models: the neoclassical, neo-Marxian, neo-Keynesian, and Kalecki-Steindl ones. In this chapter our framework has two regions, and, if each one can be made any one of the four, we have sixteen different closures! In practice, some models are logically ruled out given our framework of analysis.[6] There is, on the other hand, the

possibility that new closures not possible in a single economy one-sector framework now become possible. Instead of undertaking the tedious task of discussing each possible closure, we will examine some cases which have been discussed in the literature, and some others which are empirically 'plausible'. Most contributions make the South neo-Marxian, drawing on the analysis of Lewis (1954). Of those which do, Findlay's models make the North neoclassical, those of Taylor and Dutt, neo-Kaleckian, and that of Vines, neo-Marxian; we will examine these and also develop a model with a neo-Keynesian North. To examine the implications of giving up the neo-Marxian closure for the South, we shall examine a closure with a neo-Marxian North and a neo-Keynesian South. To illustrate how new closures become possible we shall also examine a model with a neo-Kaleckian North, but a South which does not correspond completely to any of the four categories.

4   Although this is a continuation of our second point, our models assume complete specialization. To economists brought up on HOS models this assumption will seem particularly strong. A casual glance at North-South exports may also suggest that there are few goods in which there is complete specialization.[7] We make the assumption primarily because it has been made explicitly and implicitly by most uneven development theorists, and because it makes the models very simple. Not all of the arguments made by the uneven development theorists would hold water if we did not make the assumption. One could wonder if this does not imply that the arguments which require such an assumption should be discarded, since the assumption is clearly unrealistic. We do not think so. First, casual empiricism based on broad classifications of commodities should not blind us to the fact that Northern and Southern goods are usually distinct (cars, clothing, processed foods) even when included in the same classificatory category.[8] Secondly, the point is not that our assumption makes it impossible to have incomplete specialization even in principle; it simply does not occur in the time span for which our results are relevant, that is, for which the North remains the North and the South the South. Thirdly, as already noted, some of the implications of the models are left unchanged by modifications allowing for incomplete specialization. Where they are not, it is often because the incomplete specialization models (along HOS lines) stress intersectoral efficiency issues rather than within-sector efficiency, capital accumulation and technological change; it is not clear that this emphasis is justified.

The rest of the chapter proceeds as follows. Section 8.1 examines the general framework abstracting from capital mobility and section 8.2 examines the alternative models. Section 8.3 offers some general comments on the relevance of the models to the literature on uneven development. Section 8.4 shows how capital mobility can be introduced into the analysis.

8.1 The general framework

We assume that there are two regions, a rich North and a poor South, which each produce a single good – the N good and the S good – respectively. Each good is produced with Leontief technology, using two homogeneous factors of production, capital and labour. The S good is only a consumption good, but the N good is a consumption good and an investment good (both in the North and the South). These assumptions imply that following our earlier discussions we can write two quantity equations and two price equations, one for each region, as follows:

$$X_n = c_n^n L_n + c_n^s L_s + g^n K_n + g^s K_s, \tag{8.1}$$

$$X_s = c_s^n L_n + c_s^s L_s, \tag{8.2}$$

$$P_n = W_n a_0^n + r_n P_n (K_n/X_n), \tag{8.3}$$

$$P_s = W_s a_0^s + r_s P_n (K_s/X_s), \tag{8.4}$$

with

$$K_i/X_i \geqslant a_1^i, \quad i = n,s, \tag{8.5}$$

where indices n and s denote North and South, and c_j^i denotes the consumption of good j in region i per worker employed in region i, and the other symbols have the same meanings as before. Prices are measured in terms of a common currency (with the exchange rate fixed and set equal to one); this assumption is irrelevant for models determining only relative prices.

We will continue to assume that workers do not save and receive only wage income and capitalists only earn profit income. In the North, capitalists save a constant fraction s_n of their income; the combined consumption expenditure of workers and capitalists is then split up between the Northern and Southern goods, with a constant fraction, a, being spent on the N good. In the South workers only consume the Southern good; capitalists save a constant fraction s_s and spend a constant fraction, b, of their consumption expenditure on the N good, the rest going to the Southern good. Labour and capital are internationally immobile. These assumptions imply that we can write

$$c_n^n L_n P_n = a[W_n L_n + (1 - s_n) r_n P_n K_n], \tag{8.6}$$

$$c_s^n L_n P_s = (1 - a)[W_n L_n + (1 - s_n) r_n P_n K_n], \tag{8.7}$$

$$c_s^s L_s P_s = W_s L_s + (1 - s_s)(1 - b) r_s P_n K_s, \tag{8.8}$$

$$c_n^s L_s P_n = (1 - s_s) b r_s P_n K_s. \tag{8.9}$$

Substitution of (8.6) through (8.9) in (8.1) through (8.4) implies

$$1 = a[(W_n/P_n) a_0^n + (1 - s_n) r_n (K_n/X_n)] \\ + [(1 - s_s) b r_s + g^s](K_n/X_n) k^{-1} + g^n (K_n/X_n), \tag{8.10}$$

$$1 = (1-a)\left[(W_n/P_n)a_0^n(X_n/K_n)(K_s/X_s)k\pi^{-1}\right.$$
$$+ (1-s_n)r_n(K_s/X_s)k\pi^{-1}\left] + (W_s/P_s)a_0^s\right.$$
$$+ (1-s_s)(1-b)r_s(K_s/X_s)\pi^{-1}, \tag{8.11}$$

$$1 = (W_n/P_n)a_0^n + r_n(K_n/X_n), \tag{8.12}$$

$$1 = (W_s/P_s)a_0^s + r_s(K_s/X_s)\pi^{-1}, \tag{8.13}$$

where $k = K_n/K_s$ and $\pi = P_s/P_n$.[9]

Without any capital flows, balance of payments requires balanced trade, which implies

$$(1-a)\left[W_nL_n + (1-s_n)r_nP_nK_n\right] = g^sP_nK_s + (1-s_s)br_sP_nK_s, \tag{8.14}$$

which in turn implies

$$(1-a)\left[(W_n/P_n)a_0^n(X_n/K_n)k + (1-s_n)r_n(K_n/K_s)\right] = g^s$$
$$+ (1-s_s)br_s. \tag{8.15}$$

Substitution of (8.15) and (8.3) in (8.1) implies

$$s_nr_n = g^n, \tag{8.16}$$

while substitution of (8.15) and (8.4) in (8.2) implies

$$s_sr_s = g^s, \tag{8.17}$$

which show that with balanced trade, in each region, total income equals total expenditure, so that saving equals investment.

The framework examined so far can be represented by five independent equations, that is (8.10), (8.12), (8.13), (8.16), and (8.17).[10] However, it has ten variables, that is, W_n/P_n, W_s/P_s, r_n, r_s, g^n, g^s, K_n/X_n, K_s/X_s, k, and π.

In the short run, we may fix k, assuming given stocks of capital. But, since we are confining our attention to long-run equilibria, we obtain one more equation from the fact that we are treating k as a variable to be determined, which implies that K_n and K_s must be growing at the same rate, or that

$$g^n = g^s \tag{8.18}$$

which is the condition for long-run (steady state) equilibrium for the world economy. Note that (8.18) implies, after substituting it, (8.12), (8.16), and (8.17) in (8.10),

$$k = \left[(1-s_s)b/s_s + 1\right]\left\{(1-a)\left[(1/g(K_n/X_n)) - 1\right]\right\}^{-1}, \tag{8.19}$$

where g is the common rate of growth of capital stock.

We still need four more equations to 'close' our model. In the next section we shall consider alternative models which use alternative sets of four equations to close it.[11]

For each closure we shall examine the effects of parametric shifts on the variables of the models. We shall be interested in the effects on the rates of growth, the rates of profits, the real wages, the terms of trade, and k. Note that the real wage in the North, in terms of the true cost of living index, is given by

$$V_n = W_n/(P_n^a P_s^{1-a})$$

which can be written as

$$V_n = (W_n/P_n)(P_n/P_s)^{1-a}. \tag{8.20}$$

The real wage in the South is given by $V_s = W_s/P_s$.

8.2 Alternative closures

8.2.1 Neoclassical North with neo-Marxian South

This model assumes full capacity utilization in long-run equilibrium in both economies, as in neoclassical and neo-Marxian models, full employment growth in the North as in the neoclassical model, and a fixed real wage in the South, as in the neo-Marxian model. These assumptions imply

$$K_n/X_n = a_1^n, \tag{8.21}$$

$$K_s/X_s = a_1^s, \tag{8.22}$$

$$g^n = n, \tag{8.23}$$

$$W_s/P_s = V_s, \tag{8.24}$$

where V_s is the fixed real wage. While this fixity can be explained in terms of standard neo-Marxian arguments, for less developed economies it might make the most sense to assume it to be fixed along the lines suggested by Lewis (1954), by the average income in the subsistence sector. This follows from the assumption that this sector is not capitalistically organized, and peasants share total output equally; the average product is assumed to be constant in the sector.

Substitution of (8.22) and (8.24) into (8.13) implies

$$\pi = V_s a_0^s \pi + r_s a_1^s \tag{8.25}$$

which gives a relation between r_s and the terms of trade, π, shown by OT in diagram (b) of Figure 8.1. From (8.17) we obtain the relation between g^s and r_s shown by OS in (d), and (8.16) gives the relation between g^n and r_n shown by ON in (c). Substitution of (8.21) in (8.12) gives

$$1 = (W_n/P_n)a_0^n + r_n a_1^n \tag{8.26}$$

which gives the Northern wage-profit frontier shown as AB in (a). From (8.23), the level of g^n is fixed at rate n, and (c) determines g^{n*} and r_s^*, and r_s, g^{s*}, π^*, and $W_n/P_n)^*$ are determined by the figure as shown. We can determine k^* by

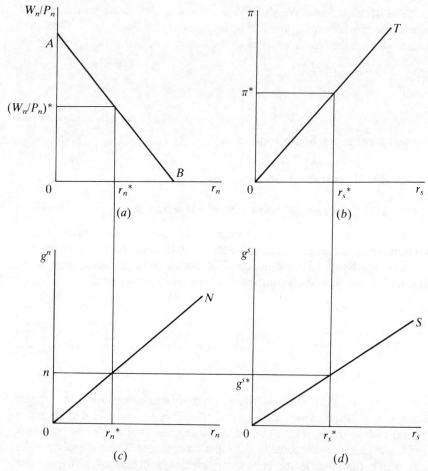

Figure 8.1   Neoclassical North and neo-Marxian South

substituting (8.21) and (8.23) in (8.10) to get

$$k^* = [(1 - s_s)b/s_s] \{(1 - a)[(1/a_1^n n) - 1]\}^{-1}. \tag{8.27}$$

V_n^*, given by (8.20), is also determined since $(W_n/P_n)^*$ and π^* are determined.[12]

We now examine the effects of variations in parameters, by examining how the curves of the figure shift and from (8.20) and (8.27). A rise in n increases g^n, leaving the curves unchanged. A rise in V_s rotates the OT curve upwards. A rise in s_n and s_s rotates ON and OS upwards, respectively; the shift in s_s also affects (8.27). Changes in a and b will not affect the figure, but will affect k^* through (8.27); the change in a is not relevant in studying the impact on the Northern real wage, which should be measured in terms of a given cost of living index. Technological changes in the South, shown by declines in a_0^s and a_1^s rotate OT

Table 8.1. *Effects of parametric changes in the model with neoclassical North and neo-Marxian South*

	g^n	g^s	r_n	r_s	W_n/P_n	V_n	V_s	π	k
n	+	+	+	+	−	−	0	+	+
V_s	0	0	0	0	0	0	+	+	0
s_n	0	0	−	0	+	+	0	0	0
s_s	0	0	0	−	0	−	0	−	−
a	0	0	0	0	0	0	0	0	+
b	0	0	0	0	0	0	0	0	+
a_0^s	0	0	0	0	0	+	0	−	0
a_1^s	0	0	0	0	0	+	0	−	0
a_0^n	0	0	0	0	+	+	0	0	0
a_1^n	0	0	0	0	+	+	0	0	−

downwards. Technological changes in the North shift the AB curve outwards: when a_0^n falls the curve rotates anchored at B; when a_1^n falls it rotates anchored at A and affects (8.27). The effects of all these changes are shown in Table 8.1, where increases are shown with '+', decreases with '−', and no effects by '0'. The change in all parameters is assumed to be in the positive direction except for the technological ones, which are assumed to fall.

Among the North-South models available in the literature this one resembles most Findlay's (1980, 1981). Findlay's model, like this one, assumes full employment growth in the North at a given rate, and a fixed real wage in the South. However, there are several differences: his allows for factor substitution, while ours assumes fixed coefficients; his, not distinguishing between Northern classes, assumes the same saving behaviour for all Northerners, while ours allows for two classes with different saving propensities. Thus, if his is a Solow-Lewis model, ours may be called a Kaldor/ Pasinetti-Lewis model. Notice that in our model without factor substitution, if we do not allow for differences in saving patterns among Northern classes, no steady state is possible in general and Harrod's long-run problem emerges. The differences in our assumptions from Findlay's imply some differences in results.

8.2.2 Neo-Marxian North and neo-Marxian South

This model assumes full capacity utilization in both the North and the South, so that (8.21) and (8.22) hold. The South has a given real wage, so that (8.24) applies. For the North, too, we assume a given real wage in the North, so that

$$W_n/P_n = V_n(P_s/P_n)^{1-a} \tag{8.28}$$

is assumed with a given V_n, perhaps for neo-Marxian reasons.

Using equation (8.22) in (8.13) implies (8.25) as in the previous model; this gives the OT line in (b) in Figure 8.2 as in the previous figure. From (8.16) through (8.18) we get

$$r_n = (s_s/s_n)r_s \qquad (8.29)$$

which gives curve OR in (d). Substitution of (8.21) in (8.12) again gives (8.26), which gives line AB in (c) as before. In (a) ON plots (8.28) for a given V_n. Curve CD in (b) is derived from curves OR, AB, and ON, and shows combinations of the terms of trade, π, and r_s which satisfy (8.26), (8.28), and (8.29). The intersection of OT and CD in (b) determines r_n^* and π^*, and r_s^* and $(W_n/P_n)^*$ can be read off from the rest of the figure.

Substitution of the rates of profit in (8.16) or (8.17) determine g^*, and

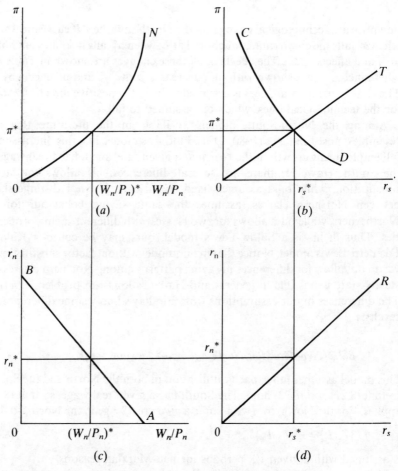

Figure 8.2   Neo-Marxian North and neo-Marxian South

substitution of this and (8.21) in (8.19) implies

$$k^* = [(1-s_s)b/s_s + 1] \{(1-a)[(1/g^*a_1^n)-1]\}^{-1}. \tag{8.30}$$

The effects of parametric shifts can be examined from Figure 8.2 and (8.16) or (8.17) and (8.30). A rise in V_n pushes the ON curve to the left, and this implies that CD is pushed to the left; the rates of profit fall, which implies, from (8.16) or (8.17), that g^* falls so that, from (8.30), k^* falls. The effects of the other parametric variations can be studied in the same way, and are shown in Table 8.2.[13] To verify these, note the following. A rise in V_s rotates OT upwards, leaving CD unchanged. A rise in s_n rotates OR downwards, shifting CD to the right, while a rise in s_s rotates OR upwards, shifting CD to the left; since g^* rises and so does s_s, from (8.30) it is not possible to sign the effect on k^*. A rise in a twists the ON curve of the fourth quadrant, pushing it leftwards above $\pi = 1$, rightwards below it, and unchanged at $\pi = 1$. The effects in the table are shown assuming an initial equilibrium $P_s/P_n = 1$. Reductions in a_0^s and a_1^s rotate OT downwards; reductions in a_0^n and a_1^n rotate AB and shift CD to the right.

Among the North-South models to be found in the literature, this model comes closest to Vines's (1984), which formalizes Kaldor (1979). However, Vines assumes that the capital-output ratio in the South depends on the amount of land, and that there is diminishing returns to land; rent income thus emerges as a third category of income in the South, and this is the source of the differences in his long-run results from ours. Thirlwall (1986) considers a closed economy model (with two sectors – an agricultural and a manufacturing) which in essence is the same as the model considered here, with differences due to the fact that he ignores the distinction between classes in the agricultural sector (our South), fixes the real wage in terms of the agricultural good ($W_n/P_n = $ constant, in our notation) and assumes $s_n = 1$. Conway and Darity (1985) consider neo-Marxian fixed wage assumptions for what they call their short run and intermediate runs; but for their long run they revert to neoclassical assumptions.

Table 8.2. *Effects of parametric changes in the model with neo-Marxian North and neo-Marxian South*

	g^n	g^s	r_n	r_s	W_n/P_n	V_n	V_s	π	k
V_n	−	−	−	−	+	+	0	−	−
V_s	−	−	−	−	+	0	+	+	−
s_n	+	+	−	+	+	0	0	+	+
s_s	+	+	+	−	−	0	0	−	?
a	0	0	0	0	0	0	0	0	+
b	0	0	0	0	0	0	0	0	+
a_0^s	+	+	+	+	−	0	0	−	+
a_1^s	+	+	+	+	−	0	0	−	+
a_0^n	+	+	+	+	+	0	0	+	+
a_1^n	+	+	+	+	+	0	0	+	?

8.2.3 Neo-Keynesian North and neo-Marxian South

This model assumes full capacity utilization in both regions, so that (8.21) and (8.22) hold. The South is still neo-Marxian, so that we assume (8.24), but the North is neo-Keynesian so that we introduce the desired accumulation function

$$g^n = g^n(r_n), \tag{8.31}$$

where, as in chapter 2, $g^n(0) > 0$, $g^{n\prime} > 0$ and $g^{n\prime\prime} < 0$.

In Figure 8.3, in (b) ON and GG represent, respectively, (8.16) and (8.31), and their intersection solves for g^{n*} and r_n^*. Then (d), where AB is obtained from

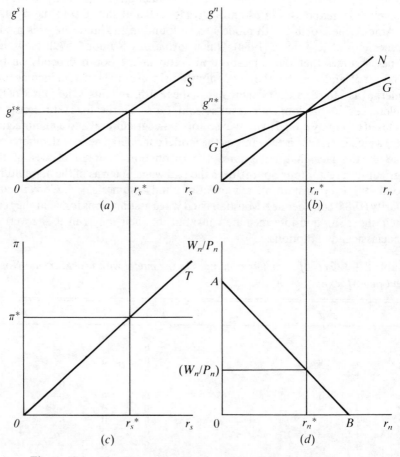

Figure 8.3   Neo-Keynesian North and neo-Marxian South

Table 8.3. *Effects of parametric changes in the model with neo-Keynesian North and neo-Marxian South*

	g^n	g^s	r_n	r_s	W_n/P_n	V_n	V_s	π	k
g	+	+	+	+	−	−	0	+	+
s_n	−	−	−	−	+	+	0	−	−
s_s	0	0	0	−	0	+	0	−	−
V_s	0	0	0	0	0	−	+	+	0
a	0	0	0	0	0	0	0	0	+
b	0	0	0	0	0	0	0	0	+
a_0^s	0	0	0	0	0	+	0	−	0
a_1^s	0	0	0	0	0	+	0	−	0
a_0^n	0	0	0	0	0	+	0	0	0
a_1^n	0	0	0	0	0	+	0	0	−

(8.21) and (8.12), solves for $(W_n/P_n)^*$; (a), where OS depicts (8.17) determines g^{s*} and r_s^* (8.17); and (c), where OT is given by (8.25), which must hold in this model, solves for π^*. Finally, k^* and V_n^* are determined from (8.30) and (8.20).

It is a simple matter now to fill in the cells of Table 8.3 as before. The only new exogenous change considered is an upward shift in the desired accumulation function, denoted by a rise in g. V_n is now a variable so that parametric variations in it cannot be considered. Finally, shifts in n are irrelevant.

While this model has not yet been examined in the literature on North-South models, it should be appealing to neo-Keynesians.

8.2.4 Kalecki-Steindl North and neo-Marxian South

This model assumes full capacity utilization in the South, so that (8.22) is assumed, but allows for the existence of excess capacity in the North, so that (8.21) is *not* assumed; it is assumed that in (8.5) the strict inequality applies for $i = n$. We thus require three equations to close the model. The first comes from the assumption of a neo-Marxian South, which implies (8.24). The second is the Kalecki-Steindl desired accumulation function

$$g^n = g^n(r_n, X_n/K_n) \tag{8.32}$$

with both partials positive, as in chapter 2. The third is the Kaleckian markup pricing equation

$$P_n = W_n a_0^n(1 + z), \tag{8.33}$$

where z is fixed.

Substitution of (8.33) in (8.12) implies, as in closed economy models

$$r_n = [z/(1 + z)](X_n/K_n) \tag{8.34}$$

which when substituted in (8.32) implies that g^n is a rising function of r_n, which we assume to be concave; it is drawn as GG in (b) in Figure 8.4. Its intersection

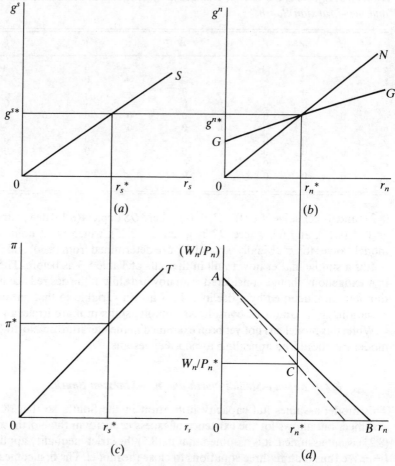

Figure 8.4   Kalecki-Steindl North and neo-Marxian South

with ON, which represents (8.16), determines g^{n*} and r_n^*. In (d) AB is the wage-profit frontier of the earlier models. Here, as discussed in chapter 2, because of excess capacity the economy must be inside it. From (8.33) we solve for $W_n/P_n = 1/(1+z)a_0^n$, so that the economy must be at point C; since C is inside AB, the inequality in (8.5) for $i = n$ is satisfied. OS in (a) shows (8.17) and solves for r_s^* and g^{s*}. OT in (c) represents (8.25) and solves for π^*. Substituting for r_n^* in (8.34) determines $(X_n/K_n)^*$. Substituting this, and the solved value of the common rate of growth, solves k^*. Finally, V_n^* is obtained from (8.20).

The long-run effects of parametric variations are shown in Table 8.4. Notice that a rise in z, shifts down the GG curve in Figure 8.4. To obtain explicit solutions for the variables we assume a linear form of (8.32) given by

$$g^n = \alpha + \beta r_n + \tau(X_n/K_n), \tag{8.35}$$

Table 8.4. *Effects of parametric changes in the model with Kalecki-Steindl North and neo-Marxian South*

	g^n	g^s	r_n	r_s	W_n/P_n	V_n	V_s	X_n/K_n	π	k
g	+	+	+	+	0	−	0	+	+	0
s_n	−	−	−	−	0	+	0	−	−	+
s_s	0	0	0	−	0	+	0	0	−	−
z	−	−	−	−	−	?	0	−	−	+
a	0	0	0	0	0	0	0	0	0	+
b	0	0	0	0	0	0	0	0	0	+
V_s	0	0	0	0	0	−	+	0	+	0
a_0^s	0	0	0	0	0	+	0	0	−	0
a_1^s	0	0	0	0	0	+	0	0	−	0
a_0^n	0	0	0	0	+	+	0	0	0	0
a_1^n	0	0	0	0	0	0	0	0	0	0

where α, β and τ are positive constants. This implies, using (8.19), that

$$k^* = s_n[(1-s_s)b/s_s + 1]\{(1-a)[1/z + (1-s_n)]\}^{-1} \qquad (8.36)$$

which we use to find the effects of changes in some of the parameters.

This model resembles those of Taylor (1981, 1983) and Dutt (1988a), although Taylor's model differs by introducing a third region, OPEC, letting the North import 'oil' as an intermediate input, and considering capital flows.

8.2.5 Kalecki-Steindl North and South with excess capacity

While the last model may be based on the assumption of competitive conditions in the S-good market, we now consider a model which allows for the existence of monopoly power in the South as well, with firms setting the price as a markup on prime costs and operating with excess capacity. Southern firms, however, do not have a desired accumulation function, but invest all savings.[14] The South therefore does not fit into any of the four closures discussed above. The North, however, satisfies the Kalecki-Steindl assumptions.

For the North we assume (8.32) and (8.33), although we denote the Northern markup by z_n to distinguish it from the Southern one. For the South we forsake (8.22) and assume markup pricing, so that

$$P_s = W_s a_0^s (1 + z_s), \qquad (8.37)$$

where z_s is the fixed markup in the South. To determine the levels of P_n and P_s, and hence π, we assume given levels of W_n and W_s.

In Figure 8.5, (b) is the same as that in Figure 8.4, and r_n^* and g^{n*} are determined in it. Also (a) and (d) are the same: g^{s*} and r_s^* are determined in (a), and the value of $(W_n/P_n)^*$ found from (8.33) shows in (d) at which point inside the potential wage-profit frontier AB the Northern economy will be at equilibrium. In (c) FG shows the potential wage-profit frontier for the South.

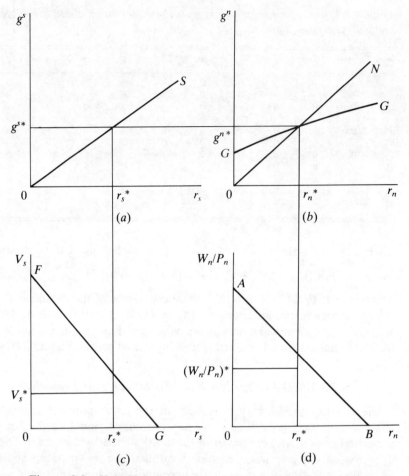

Figure 8.5   Kalecki-Steindl North and South with excess capacity

With $(W_s/P_s)^* = 1/a_0^s(1+z_s)$ found from (8.37), the point within the frontier that the South will be at is determined; operation within the frontier ensures the existence of excess capacity there. The Southern rate of profit is given by

$$r_s = [W_s z_s a_0^s/(1+z_n)W_n a_0^n](X_s/K_s) \tag{8.38}$$

which can solve for $(X_s/K_s)^*$, the degree of Southern capacity utilization. From (8.33) and (8.37)

$$\pi^* = W_s a_0^s(1+z_s)/W_n a_0^n(1+z_n). \tag{8.39}$$

Assuming the linear investment function given by (8.35) we can also show that

$$k^* = s_n[(1-s_s)b/s_s+1]\{(1-a)[1/z_n+(1-s_n)]\}^{-1}. \tag{8.40}$$

Finally, the Northern real wage can be found to be, from (8.20) and (8.39),

$$V_n = W_n^{1-a}/[a_0^n(1+z_n)]^a[W_s a_0^s(1+z_s)]^{1-a}. \tag{8.41}$$

The effects of parametric variations are shown in Table 8.5. Here the money wages become parameters. The model here is the same as the one developed more fully in Dutt (1984b).

8.2.6 Neo-Marxian North and neo-Keynesian South

All the models discussed in this chapter, with the sole exception of the previous one, have a neo-Marxian South. To explore the significance of that assumption, we now consider a model with a neo-Marxian North and a neo-Keynesian South: the exact opposite of the model of section 8.2.3.

We thus assume (8.21), (8.22), and (8.28) with a fixed V_n, and introduce a Southern desired accumulation function,

$$g^s = g^s(r_s), \tag{8.42}$$

where $g^{s\prime} > 0$, $g^{s\prime\prime} < 0$ and $g^s(0) > 0$.

In Figure 8.6 GG, the Southern desired accumulation function given by (8.42), and OS given by (8.17), determine g^{s*} and r_s^*. ON, representing (8.16), determines r_n^*. AB, the Northern potential wage-profit frontier determines $(W_n/P_n)^*$ and OT, representing (8.28) determines π^*. FG, the Southern wage-profit frontier determines V_s^*, once its slope has been fixed after the terms of trade are known. From (8.30) we determine k^*, once g^*, the common growth rate, is determined in the figure.

The effects of parametric shifts are shown in Table 8.6. It should be

Table 8.5. *Effects of parametric changes in the model with Kalecki-Steindl North and South with excess capacity*

	g^n	g^s	r_n	r_s	W_n/P_n	V_n	V_s	X_n/K_n	X_s/K_s	π	k
g	+	+	+	+	0	−	0	+	+	0	0
s_n	−	−	−	−	0	0	0	−	−	0	+
s_s	0	0	0	−	0	0	0	0	−	0	−
z_n	−	−	−	−	−	−	0	−	−	−	+
z_s	0	0	0	0	0	−	−	0	−	+	0
W_n	0	0	0	0	0	+	0	0	+	−	0
W_s	0	0	0	0	0	−	+	0	−	+	0
a	0	0	0	0	0	0	0	0	0	0	+
b	0	0	0	0	0	0	0	0	0	0	+
a_0^s	0	0	0	0	0	0	+	0	+	−	0
a_1^s	0	0	0	0	0	0	0	0	0	0	0
a_0^n	0	0	0	0	+	+	0	0	−	+	0
a_1^n	0	0	0	0	0	0	0	0	0	0	0

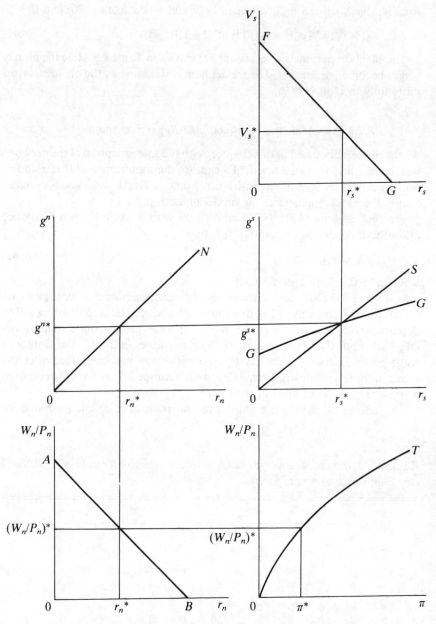

Figure 8.6   Neo-Marxian North and neo-Keynesian South

Table 8.6. *Effects of parametric changes in the model with neo-Marxian North and neo-Keynesian South*

	g^n	g^s	r_n	r_s	W_n/P_n	V_n	V_s	π	k
g	+	+	+	+	−	0	−	−	+
s_n	0	0	−	0	+	0	0	+	0
s_s	−	−	−	−	+	0	+	+	−
V_n	0	0	0	0	0	+	0	−	0
a	0	0	0	0	0	0	0	0	+
b	0	0	0	0	0	0	0	0	+
a_0^s	0	0	0	0	0	0	+	0	0
a_1^s	0	0	0	0	0	0	+	0	0
a_0^n	0	0	0	0	+	0	0	+	0
a_1^n	0	0	0	0	+	0	0	+	−

remembered that shifts in π will change the slope of the Southern wage-profit frontier – improvements in the terms of trade are analogous to increases in the productivity of capital. The rise in g refers to shifts in the desired accumulation curve for the South. Among the results, notice in particular that the effects of technological progress are very different from what they are in the other models of this section.

8.3 Comments on uneven development

Each of the models of section 8.2 found the long-run equilibrium values of the variables of the model, including k^*, given the parameters of the model. They also showed how the variables, including k^*, change due to parametric shifts. We may define uneven development as a rise in k^* due to a parametric shift,[15] and call the parametric shift causing this rise 'a valid mechanism of uneven development'. Note that a rise in k^* implies that, in the dynamic path from one long-run equilibrium to another (on both of which the North and the South have equal growth rates), the North, on average, must grow (in capital stock) more than the South.

It may be objected that it is inappropriate to study the phenomenon of uneven development by examining comparative dynamic effects which compare long-run equilibria in which North and South grow at the same rate. It can be argued that uneven development implies an unstable cumulative process which does not lead to a stable long-run equilibrium (as modelled in Krugman, 1981a, and Dutt, 1986a, for example); or that the long-run equilibrium should allow the North and South to grow at divergent rates (as in Conway and Darity, 1985). But simplicity is our method's defence. Further, the endogenization of some of the long-run parameters of the models can yield cumulative processes in the context of broader models, as will be seen in later chapters. The North and South can also be made to grow at divergent rates in long-run equilibrium, but that is not possible within our general framework.

While several possible mechanisms of uneven development are examined in the informal literature, the discussions are seldom explicit about the theoretical framework assumed. It is therefore of interest to introduce these mechanisms in our models in the form of parametric shifts. We confine our attention to only a few well-known examples.

1   Singer (1975) and Prebisch (1963) and others examine shifts in Northern expenditure which increase the proportion of consumption expenditure spent on Northern goods. In our models such shifts can be represented as increases in a. All our models show that this is a valid mechanism of uneven development, so that its validity does not depend on the specific theoretical framework adopted. Chapter 9 will present a fuller exposition of this mechanism.

2   Singer and Prebisch also examine the implications of technological change. Improvements in the North are supposed to be advantageous to the North, while the benefits of improvements in the South are siphoned off to the North. This is a claim that seems to be valid for several models. In those that allow the endogenous determination of the Northern real wage (that is, with the exception of models with a neo-Marxian North), technological changes everywhere raise the Northern real wages, while neo-Marxian assumptions, however, keep the Southern real wage fixed. Notice that this has nothing to do with product market imperfections for the Northern good: it happens also in models with perfectly competitive N goods markets. Notice also that in these models technological change does not result in uneven development (in fact in a few cases it results in even development). In the model with fixed real wages in both economies, technological changes everywhere typically are valid mechanisms for uneven development. In the model with a neo-Marxian North and a neo-Keynesian South, technological change in the South actually raises real wages there; this again underscores the importance of a neo-Marxian South for the Singer-Prebisch result.

3   Baran (1957) examines the implications of a rise of monopoly power in the North for international development. Only the two models which have a Kalecki-Steindl closure for the North can address this issue, and they both show that a rise in monopoly power in the North, as reflected in a rise in z (or z_n) results in uneven development. Baran's argument, of course, is far more complex, involving Northern tendencies towards stagnation, the importance of military spending to reverse this tendency, which is in turn related to the imperialistic domination of the South by the North, which through a variety of mechanisms causes Southern deindustrialization. Our formulation presents a more rigorous analysis of the effects of the rise of monopoly power on Northern growth than is available in the work of Baran or Sweezy, but ignores the discussion of imperialism (except to the

extent that it determines the structure of the world economy). Uneven development in our models is only a result of changes in demand composition. The increase in monopoly power in the North, by redistributing income towards capitalists and workers, reduces aggregate demand in the North and results in stagnation, as in our closed economy Kalecki-Steindl models described before. The demand for the Southern good falls both because of Northern stagnation, and changes in the pattern of spending in the North away from Southern goods as a result of greater proportional spending on the investment good. The Southern terms of trade thus deteriorates, and adversely affects its rate of growth.[16]

4　Emmanuel (1972) examines the implications of a rise in the Northern real wage. To consider the effects of an exogenous increase in the Northern real wage, we need a model which makes it a parameter. The model with the neo-Marxian North and South implies that a rise in the Northern real wage reduces the Southern terms of trade and the rates of growth of both economies, but results in even development. The model with the neo-Marxian North and the neo-Keynesian South does not affect growth rates anywhere, worsens the Southern terms of trade, but does not result in uneven development. These are hardly what Emmanuel would admit as formalizations of his argument, that a rise in the Northern real wage increases demand and rates of accumulation in the North. The models with the Kalecki-Steindl North allow for this possibility. If a strengthening of the position of workers in the 'class struggle' reduces z (or z_n), this will raise the rate of Northern growth. The reduction, however, will improve the Southern terms of trade and result in even development. To allow for uneven development in these models we could make a rise as the North grows faster when z falls.[17]

Emmanuel's work has traditionally been formalized with the Sraffian prices of production approach.[18] If we take a simple Sraffa-type model with given coefficients of production, assume internationally equalized profit rates (due to perfect capital mobility), and take given and equal wages in both regions, we can determine the implied terms of trade, and call it equal exchange. If the Northern wage is then made higher than that in the South, the terms of trade can be shown to favour the North compared to the case of equal exchange. Emmanuel defines this as unequal exchange, and treats it as a transfer of surplus. Such a comparison of the actual terms of trade with the hypothetical equal exchange case cannot prove that the South is getting impoverished as a result of this transfer. Any terms of trade more favourable to the South in comparison to an arbitrary reference level will imply a surplus transfer which slows down Southern growth. Only an ethical case can be made on the basis of the 'fair' equal exchange situation, but Emmanuel clearly wants to do more: unequal exchange to him is a mechanism of uneven development.

Our approach stresses another part of Emmanuel's argument, which claims that, because the North is more developed, its wage rises through time, and this results in the faster development of the North. This dynamic argument states that the higher wage in the North increases demand and growth of the North, and can be formalized by our Kalecki-Steindl model for the North. But this is also the model in which the input-output ratios in Sraffa's model cannot be taken to be fixed. As we saw in chapters 6 and 7, Sraffa's given input-output ratios imply an inverse relation between wages and profit rates: higher wages will reduce the rate of profit and hence Northern growth, contrary to Emmanuel's argument that higher wages boosts the rate of growth, which requires (in the absence of technological changes) a higher degree of capacity utilization, and hence a Kalecki-Steindl model. The Sraffian approach to Emmanuel thus needs to be replaced by a Kalecki-Steindl one. The Sraffian approach, in any case, is used to illustrate the idea of unequal exchange, which we have found to be wanting.

Our analysis also points to Samuelson's (1973, 1975, 1976, 1978) failure to point out the illogic of Emmanuel's analysis. Samuelson argues that, while with different rates of profit between regions a deadweight loss is possible because regions can (in steady state) end up having the 'wrong' kind of specialization, this cannot be for Emmanuel's case with equalized rates of profit. Samuelson's analysis is flawed because he uses a neoclassical model with full employment of labour at steady state, with the rate of profit presumably given by time preference and wages determined by supply and demand; in Emmanuel's system, wages in the two countries are exogenously given and the rate of profit is determined endogenously (see Emmanuel, 1978). Optimality propositions derived from neoclassical models do not carry over to models with other types of 'closures'.

5   Braun (1984) explores the consequences of restrictions on the imports of Southern goods into the North. Braun's own approach is to take a Sraffian system with complete specialization, and to show that manipulation of the terms of trade by the North results in unequal exchange in Emmanuel's sense: wages are lower in the South than in the North. We have already argued against the notion of unequal exchange, but our models support what seems to be Braun's central idea. Protection imposed by the North against Southern goods switches Northern consumption expenditure away from the Southern good and towards the Northern good, that is, raises a. This, as we have seen, worsens the Southern terms of trade in the long run, and results in uneven development.

6   There has been much discussion of the secular decline of the Southern terms of trade.[19] Our analysis gives some support to the view that the controversy over the terms of trade debate is misplaced. In a model in which the Southern terms of trade is endogenously determined, there is no presumption that its deterioration implies Southern stagnation or uneven development. In some of our models we in fact find an inverse correlation

between terms of trade deterioration and uneven development. For the model with the neoclassical North and the neo-Marxian South, for instance, there is no parametric shift which leads to both terms of trade deterioration and uneven development. However, in several models, and for some important mechanisms, there appears to be a direct correlation.

8.4 Capital mobility in North-South models

Our analysis has so far abstracted from the international mobility of capital and labour. While international labour mobility is severely restricted, at least in the modern world, the same cannot be said of capital. International capital flows – between the North and the South – have been significant for a long time, and its effects have attracted much attention in the informal literature on uneven development. This section examines how our general framework can be extended to allow for capital flows from the North to the South. In doing so, we shall maintain all assumptions made in this chapter so far, but in addition allow Northern capitalists to own a part of the capital of the South, and to earn profits from it. As before, we can take the ratio of capital stocks installed in the North and the South as an indicator of uneven development. Additionally, we now have another ratio of some significance, which measures the share of Southern capital owned by Northern capitalists to total Southern capital. This ratio can be taken as an indicator of the penetration or 'domination' by Northern capital in the South.

Given our assumptions, (8.10) and (8.11) have to be replaced by

$$1 = a\{(W_n/P_n)a_0^n + (1-s_n)[r_n(K_n/X_n)$$
$$+ r_s(K_f/K_s)(K_s/K_n)(K_n/X_n)]\}$$
$$+ (1-s_s)br_s(K_d/K_s)(K_s/K_n)(K_n/X_n) + g^n K_n/X_n$$
$$+ g^s(K_s/K_n)(K_n/X_n) \tag{8.43}$$

and

$$1 = (1-a)\{(W_n/P_n)a_0^n + (1-s_n)[r_n(K_n/X_n)$$
$$+ r_s(K_f/K_s)(K_s/K_n)(K_n/X_n)]\}(P_n/P_s)$$
$$+ (1-b)(1-s_s)r_s(K_d/K_s)(K_s/K_n)(K_n/X_n)$$
$$+ (W_s/P_s)(K_s/X_s) \tag{8.44}$$

where K_f and K_d denote Southern capital owned, respectively, by Southern and Northern capitalists, and $K_s = K_d + K_f$. The price equations are given as before by (8.12) and (8.13).

With capital flows and profit repatriation accounted for, the balance of payments can be written as

$$P_n g^s K_s + br_s P_n K_d(1-s_s) - (1-a)[W_n a_0^n X_n$$
$$+ (1-s_n)(r_n P_n K_n + r_s P_n K_f)] + r_s P_n K_f - P_n g^f K_f = 0, \tag{8.45}$$

where g^f is the rate of growth of foreign capital in the South.

Substitution of equation (8.45) in (8.43) and (8.44), using (8.12) and (8.13), implies

$$g^n k + g^f f - s_n(r_n k + r_s f) = 0 \tag{8.46}$$

and

$$g^s - s_s(1-f)r_s - g^f f = 0, \tag{8.47}$$

where $k = K_n/K_s$ and $f = K_f/K_s$. These replace (8.16) and (8.17), which did not allow for capital flows. Also, (8.44) can be rewritten, using (8.12) and (8.13), as

$$(1-a)\left[(X_n/K_n)k + r_s f - s_n(r_n k + r_s f)\right]$$
$$-\{f + [s_s + b(1-s_s)](1-f)\}r_s = 0. \tag{8.48}$$

With this equation clearing the market for the Southern good, the other equations of our model will ensure that by Walras's law, the market for the Northern good will also clear, that is, (8.43) will also be satisfied.

Our model can now be described by the five equations (8.12), (8.13), and (8.46) through (8.48). However, our model has twelve variables: g^n, g^s, g^f, r_n, r_s, W_n/P_n, W_s/P_s, π, X_n/K_n, X_s/K_s, k, and f. We need seven more equations to solve our model (together with additional existence conditions to yield economically meaningful solutions).

Focusing on long-run equilibrium as before, we obtain two more equations,[20] (8.18) and

$$g^n = g^f. \tag{8.49}$$

We now need five more equations to 'close' our model. Four of these can be found from the specific characteristics of the Northern and Southern economies, as discussed in section 8.2. A remaining one can be obtained by making a hypothesis about the determinants of g^f. A possible candidate can be

$$g^f = g^n + \tau(r_s - r_n) \tag{8.50}$$

with $\tau < 1$, which suggests that g^f will be higher than g^n when the rate of profit in the South exceeds that in the North. Notice that (8.49) implies that in equilibrium, the rates of profit are internationally equalized. Other possibilities also exist, as will be mentioned below.

With the same number of equations as variables, we can try to solve for the long-run equilibrium values of the variables in the same manner as in the case of no capital mobility. We can also examine the questions of the stability of the equilibria by exploring short-run equilibria. Finally, we can explore the consequences of exogenous changes, both for k^* (denoting uneven development) and f^* (denoting foreign domination), the long-run equilibrium values. Rather than pursue the many possibilities here, we confine ourselves to only three remarks.

First, with capital mobility determined by equation (8.50), there will in general be no interior long-run equilibrium with $0 < f < 1$; the world economy will tend to an extreme, with either total foreign domination of the South or with no foreign ownership of Southern capital.

This can be seen as follows. At long-run equilibrium, (8.18) and (8.49) imply that all three growth rates are equal, to say g. Further, the rates of profit will be equal, to say r. In this case, from (8.47) it follows that

$$g = s_s r.$$

Substitution into (8.46) implies

$$gk + gf - s_n[(g/s_s)k + (g/s_s)f] = 0$$

which reduces to

$$(s_s - s_n)(k + f) = 0.$$

This implies that either the savings propensity of capitalists in the two regions are equal, or $k = -f$, which is impossible given that k and f must be positive. Since s_n cannot in general be taken to be equal to s_s, we conclude that the type of equilibrium we have assumed here does not exist, irrespective of the assumptions made about the Northern and Southern economies.

Let us explore the characteristics of equilibrium assuming that Northern capitalists have a higher propensity to save, that is, $s_n > s_s$. In this case, assume that $f = 0$ occurs at the equilibrium. Equation (8.46) implies

$$g^n = s_n r_n \tag{8.51}$$

and (8.47) implies

$$g^s = s_s r_s. \tag{8.52}$$

At equilibrium, given (8.18), this implies that $s_n r_n = s_s r_s$, which implies that $r_s > r_n$. But in this case, Northern capitalists will want to invest in the South, so that we cannot reasonably assume $f = 0$. This absurdity implies that $f = 1$.

To check whether this outcome is a consistent one, we note that in this case (8.46) implies

$$(g - s_n r)(k + 1) = 0, \tag{8.53}$$

while (8.47) implies

$$g^s = g^f. \tag{8.54}$$

Since (8.54) implies, with (8.18), (8.49), this is consistent with $r_n = r_s$. From (8.54) it follows that in equilibrium, $g = s_n r$. Thus this case is internally consistent, and implies that f tends to 1. Exact equality requires that K_d becomes zero, but this is impossible because of our assumption of non-depreciating capital.

We thus conclude that if $s_n > s_s$, in long-run equilibrium, there will be a tendency towards the complete domination of the South by Northern capital. This corroborates Mainwaring's (1980) analysis of the Pasinetti process in the context of international investment, where the group with the highest saving propensity dominates all others.

If we insist on an interior solution to the model and have an equation similar to (8.50) as well, we may replace it by

$$g^f = g^n + \tau(\Theta r_s - r_n) \tag{8.55}$$

with $0 < \Theta < 1$. This will imply that, in equilibrium, $r_s > r_n$, so that there are some barriers to capital mobility, due to uncertainty in the South, for example. In this case, with the further restrictions that $s_n > s_s$, but $s_s > \Theta s_n$, we could obtain interior equilibria, and examine their stability and comparative dynamic properties.

Secondly, in the presence of capital mobility, we can now have certain types of closures that are not possible without capital mobility. For example, we may now have independent desired accumulation functions of the neo-Keynesian type both in the North and in the South, together with full capacity utilization in both economies, and a Lewis-like fixed real wage in the South. We will then have to have g^f as being determined by the other equations, and there will be no room for an equation of the type (8.50) or (8.55).

Thirdly, we should comment on the fact that our method of dealing with capital mobility is different from those of several recent contributions to the literature. Some of these contributions, including Bacha (1978) and Burgstaller (1985), assume that capital only is a wage fund, while our analysis allows for fixed capital, which is certainly more relevant for capital mobility in non-Ricardian times. Brewer (1985) extends the notion of capital to include circulating capital as intermediate good and wage funds, but does not allow for fixed capital. Others, including Burgstaller and Saavedra-Rivano (1984), allow for fixed capital, but have taken capital to be perfectly malleable: they assume that even in the short run, capital can flow from the North to the South to equalize rates of profit internationally. We believe our assumption to be a preferable one, though one which complicates the analysis somewhat. Finally, several contributions, including Taylor (1986), Darity (1987), and Dutt (1987d), introduce asset market complications into North-South models to analyse the debt problems of the South. Darity gives careful attention to the international banking system, but generally confines attention only to the short run. Taylor and Dutt analyse the long run as well. Complications of the type introduced in these contributions can be introduced into our models as well, to deal with asset markets which are ignored in our analysis here.

Apart from these differences, however, these contributions can be thought of as being similar to particular closures of our general framework. Thus Burgstaller and Saavedra-Rivano (1984) and Burgstaller (1985) have a neo-classical North and a neo-Marxian South, Brewer (1985) considers symmetric cases in which both North and South are neo-Marxian or neoclassical, Darity (1987) considers a model which is almost the same as assuming Kalecki-Steindl assumptions for both the North and the South, Taylor (1986) has a Kalecki-Steindlian North and a neo-Marxian South, while Dutt (1987d) has a Kalecki-

Steindl North and a neo-Marxian South with Southern investment given exogenously, and capital flows responding in an endogenous manner.

8.5 Conclusion

This chapter has developed a general framework in which several different models of North-South trade can be looked upon as special cases. Some of the models developed have close relatives in the formal literature, and some others may have somewhat less close relatives (who may disown them, uncharitably perhaps!) in the informal literature. We have examined the long-run equilibrium positions of several models, and analysed the impact of exogenous changes in them. In particular, we have examined mechanisms of uneven development in the precise sense of long-run increases in the ratio of Northern to Southern capital stock.

Some caveats, however, are in order. First, the general framework adopted is a somewhat specific one. For example, a given pattern of specialization is assumed, the South is not allowed to produce investment goods, particular saving and spending patterns are postulated, and fixed coefficients of production are assumed. Our defence here is that our purpose is to show how different models can be treated as alternative closures of some general model; if one does not like the general model, suitable alterations of it can be made without damaging this method of analysis. Modifications allowing for, for example, variable coefficients – following the lines pursued in chapter 3 – would be rather simple. Secondly, we have focused only on long-run equilibria. The study of short-run phenomenon both for itself and for the study of long-run dynamics and the stability of long-run equilibrium can be examined by the specification of the short-run structure of the models in detail. Some such analyses already exist in the literature; those for other models pose no particular problems and can be pursued along the lines suggested in footnotes. In the next chapter we shall explore the dynamic behaviour of one of the models discussed in this chapter.

The central implication of this chapter is that one's view of the evolution of the international economy, and in particular, of the nature of Southern development, depends on one's view of the structure of the world economy. It is hoped that contributors to debates in this area will be more aware of their view of that structure by being able to see what exactly they are assuming about the world, and how their assumptions differ from those who view it in some other way.

CHAPTER 9

Endogenous preferences and uneven development

9.0 Introduction

The previous chapter examined several different models of trade between a rich North and a poor South and commented on several different mechanisms through which uneven development can occur. In analysing several models and mechanisms, however, we paid insufficient attention to particular models and specific mechanisms. We focused on long-run equilibrium positions and ignored short-run dynamics. Also, uneven development was considered by subjecting the long-run equilibria to parametric shifts, rather than by examining it as a process, by endogenizing the relevant parameters involved. This does not mean that the examination of the behaviour of these models in the short run, and the analysis of uneven development as processes cannot be examined by using the models. This chapter uses the model with the Kalecki-Steindl North and the neo-Marxian South to examine one set of mechanisms of uneven development in detail.

The two mechanisms we have singled out in this chapter have been widely discussed. They relate to adverse movements in the terms of trade due to the inelastic demand for goods produced by poor countries,[1] and the international demonstration effect on consumption patterns in poor countries. The mechanisms are examined together because they can both be looked upon as affecting spending patterns. We incorporate them in our formal North-South model to understand their implications for uneven development, beyond what can be obtained from the verbal discussions that abound. The model makes preferences (to use neoclassical terminology) endogenous, holding as given other parameters which can cause uneven development. This analysis is premised on the idea that it is possible to analyse the role of one mechanism without examining the process of uneven development in its entirety. Our formal approach necessitates such piecemeal theorizing; indeed, it would appear that all useful theorizing – apart from the statement that everything depends on everything else – has to be piecemeal!

The model we use was described in section 8.2.4, though there we examined only its long-run equilibrium properties. We continue assuming that there are only two regions, the North and the South, and that they are specialized completely in the production of their two goods. Further, we assume away international factor movements, and assume balanced trade. The important differences between the two countries are as follows. The South, with an underdeveloped industrial base, cannot produce investment goods, but must import machines from the North; the North only purchases consumption goods from the South.[2] The North operates like a Kalecki-Steindl economy with an excess capacity of capital, and in a manner consistent with the existence of imperfections in competition in the market for its product; the South produces at full capacity and in competitive markets for its products. Northern firms make independent investment decisions, driven by 'animal spirits', while Southern firms either invest their own savings or are limited in their investment decisions by the availability of foreign exchange. The real wage in the South is fixed at subsistence, while in the North it may respond to economic variables. While this of course is not the only way of modelling the international economy – witness the various models discussed in the previous chapter – it seems to capture some of the characteristics of the world economy as observed by many economists who are loosely called 'structuralist' and, in any case, is enough to illustrate the uses of the models of the previous chapter.

Section 9.1 presents the model, which we call our basic model, exploring its short-run equilibrium and dynamics. Section 9.2 examines the mechanisms of low income elasticity of demand and international demonstration effects, and suggests simple formalizations of them. Section 9.3 incorporates these mechanisms into the model and examines their relevance for the process of uneven development.

9.1 The basic model

Although the structure of the model is the same as the Kalecki-Steindl North-neo-Marxian South model discussed in section 8.2.4, we review it here to refresh our memory and have all the equations for ready reference.

9.1.1 Structure of the model

The model considers trade between two regions, a developed North and an underdeveloped South.

The North produces a single good, which is both an investment good and a consumption good, with a Leontief production function, using two homogeneous factors of production, capital (physically the same as the produced good) and labour. The North has excess capacity of capital, and the product price is determined by

$$P_n = (1 + z)W_n a_0^n. \tag{9.1}$$

The actual employment of labour is equal to the demand for labour, and the money wage may be taken as fixed as a first approximaton.[3] The markup income goes to capitalists who save a fraction s_n of their income, while wage income goes to workers, who consume all their income. Total consumption expenditure in the North is thus given by

$$W_n a_0^n X_n + (1 - s_n) z W_n a_0^n X_n.$$

Implicitly assuming that capitalists and workers have identical consumption patterns, a fraction a of this – assumed to be fixed for now – is spent on the Northern good and the rest on the Southern good. Northern firms add to their stock of capital according to an investment function given by

$$I_n = g^n(r_n) K_n, \tag{9.2}$$

where $g^{n'} > 0$.[4] The rate of profit is, using (9.1), given by

$$r_n = [z/(1 + z)](X_n/K_n). \tag{9.3}$$

The South also produces a single good – which is only a consumption good – with a Leontief production function using labour and capital (which is the Northern good). Production occurs at full capacity, so that

$$X_s = K_s/a_1^s. \tag{9.4}$$

The real wage in the South is fixed in terms of the Southern good, the only good consumed by Southern workers. Profit income goes to Southern capitalists who save a fraction s_s of their income; they spend a fraction b, which is fixed for now, of the remainder of their income on the Northern good, and the rest on the Southern good. Since Southern workers do not save, consumption demand in the South for the Northern and Southern goods, respectively, are then

$$b(1 - s_s)(1 - V_s a_0^s) P_s K_s/a_1^s$$

and

$$(1 - b)(1 - s_s)(1 - V_s a_0^s) P_s K_s/a_1^s + V_s a_0^s P_s K_s/a_1^s.$$

We assume that $1 > V_s a_0^s$, so that Southern capitalists can earn profits. Investment in the South is determined in a manner discussed below.

The North and the South interact with each other by trading in goods.

9.1.2 Short-run equilibrium

Short-run equilibrium in the model requires that for given values of K_n, and K_s, the markets for the Northern and Southern goods clear. The market for the Northern good is quantity cleared in a Kaleckian fashion, while that for the Southern good is cleared by its price responding positively to the excess demand for it. The short-run equilibrium conditions are thus given as

$$a[1+(1-s_n)z]W_n a_0^n X_n + b(1-s_s)(1-V_s a_0^s)P_s K_s/a_1^s$$
$$+ P_n I_n + P_n I_s - P_n X_n = 0, \tag{9.5}$$

$$(1-a)[1+(1-s_n)z]W_n a_0^n X_n + P_s V_s a_0^s K_s/a_1^s$$
$$+ (1-b)(1-s_s)(1-V_s a_0^s)P_s K_s/{}_1^s - P_s K_s/a_1^s = 0, \tag{9.6}$$

where X_n and P_s are the only variables in the short run.

Regarding the balance of payments we have already implicitly assumed a fixed exchange rate regime (with the exchange rate set equal to unity). Payments balance requires, with no capital mobility, balanced trade, so that

$$(1-a)[1+(1-s_n)z]W_n a_0^n X_n - b(1-s_s)(1-V_a a_0^s)P_s K_s/a_1^s - P_n I_s = 0. \tag{9.7}$$

With X_n and P_s being determined in (9.5) and (9.6), we need an additional degree of freedom to satisfy (9.7). With everything else fixed, we see that I_s cannot be independently specified, but must vary to satisfy it. Possible mechanisms which can achieve this will be considered later.

Substituting (9.1), (9.2), (9.3), and (9.7) in (9.5) yields

$$s_n[z/(1+z)]u = g^n(uz/(1+z)), \tag{9.8}$$

where $u = X_n/K_n$. From (9.6) we also obtain

$$(1-a)[1+(1-s_n)z]X_n[1/(1+z)]$$
$$- [s_s + (1-s_s)b](1-V_s a_0^s)(K_s/a_1^s)\pi = 0, \tag{9.9}$$

where, as before, $\pi = P_s/P_n$, the terms of trade between the Northern and the Southern good. Dividing both sides of (9.7) by P_n and comparing with (9.9) implies also that

$$I_s = s_s(1 - Vsa_0^s)(K_s/a_1^s)\pi. \tag{9.10}$$

We can now see that there are two possible mechanisms which can make I_s endogenous, so that it can satisfy the balance of payments equation (9.7). First, we may assume that Southern savings are automatically invested, due to the fact that Southern firms are owned by capitalists who invest entirely from their own profits (above what they consume). In this case, we assume (9.10) is always satisfied, which, using (9.5) and (9.6), implies (9.7). Secondly, we may assume that Southern capitalists operate with a binding foreign exchange gap which restricts their investment plans to satisfy the balance of payments restriction, the government making sure that it is not violated; this mechanism can be formalized by requiring that I_s responds positively to a Southern trade surplus. We thus assume (9.7) is satisfied and (9.10) will be satisfied as a result.[5] The equilibrium solution of the model will be the same whichever mechanism is assumed.

The equilibrium value of u is determined from (9.8), which then determines X_n^* for the given K_n; a geometrical depiction is given in Figure 9.1, where OS and GG show the left- and right-hand sides of (9.8).[6] Insertion of X_n^* in (9.9) determines π^*, as shown in Figure 9.2, where DD is the Northern demand curve for the Southern good, and SS the marketed surplus curve for the Southern good, both in S good units; if the solution for X_n exists, so will that for π.[7]

9.1.3 Long-run dynamics and equilibrium

In the long run, K_n and K_s change over time according to the level of investment in each region. Without depreciation of capital, we have (with overdots denoting time derivatives)

$$\dot{K}_n = I_n \tag{9.11}$$

and

$$\dot{K}_s = I_s. \tag{9.12}$$

To examine the long-run evolution of the system, we consider in turn the growth rates of the North and the South, g^n and g^s, which measure the growth rates of the stocks of capital. From (9.2), (9.3), and (9.11) we get

$$g^n = \hat{K}_n = g^n(uz/(1+z)). \tag{9.13}$$

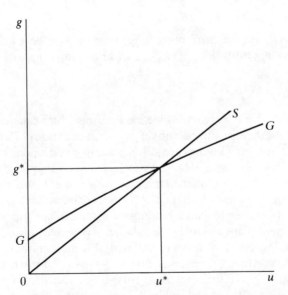

Figure 9.1   Determination of Northern capacity utilization in the short run

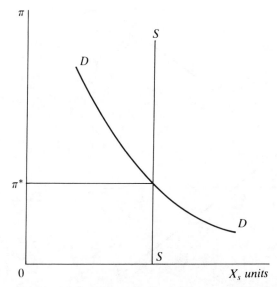

Figure 9.2   Determination of the terms of trade in the short run

Also, (9.8) shows that the equilibrium value of u depends only on the values of z and s_n and the form of the g^n function; all of this implies that the level of g^n is independent of the levels of the two state variables, K_n and K_s. For the South, on the other hand, we get from (9.10) and (9.12)

$$g^s = \hat{K}_s = s_s[(1 - V_s a_0^s)/a_1^s]\pi, \tag{9.14}$$

which shows that g^s varies positively with the terms of trade. But (9.9) implies

$$\pi = \{(1 - a)[1 + (1 - s_n)z]/(1 + z)[s_s + (1 - s_s)b][(1 - V_s a_0^s)/a_1^s]\}uk \tag{9.15}$$

which shows that π in turn, depends on k, the ratio of the stocks of capital in the North and the South. With u determined in (9.8), and all of the parameters held constant, this relation is seen to be positive. Combining (9.14) and (9.15) we find a positive, proportionally linear relationship between g^s and k.

The natural characterization of a long-run equilibrium in this framework is the state at which $g^n = g^s$. This is shown in Figure 9.3, where k^* is the long-run equilibrium value of k, which is the natural state variable of the system. The figure also shows that this long-run equilibrium is stable; starting with $k < k^*$, $g^s < g^n$, which will make k rise back to k^*. It is easy to check that the long-run stability condition is the same as the condition that g^s is upward rising, which we have already found to be the case.

An implication of the stability of this long-run equilibrium is that, if we start with a situation with $g^s > g^n$, then the gap between the rates of growth between the North and South will narrow and eventually they will become equal. In the

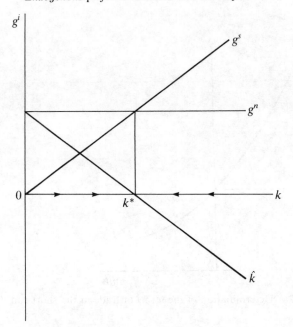

Figure 9.3   Long-run equilibrium in the basic model

context of this model it seems natural to call this pattern of international development 'even'. 'Uneven' development can then be said to occur when the development is not even. Uneven development thus requires that the gap between the rates of growth increases over time; this implies that k consistently increases over time. It should be noted that this definition of uneven development is a rather weak one in the sense that our notion of even development includes the case in which the absolute gap between the North and the South grows over time, and the ratio of growth rates in per-capita terms may move in favour of the North (given that population growth rates in the South are higher than in the North).[8]

9.2 Changes in demand composition

This section examines changes in demand composition due to the low income elasticity of Southern goods in the North, and the operation of the international demonstration effect.[9]

9.2.1 Low income elasticity of the Southern good in the North

In discussions of the secular decline of the Southern terms of trade, one factor which has received repeated attention is that concerning the decline in the share of consumer expenditure on Southern goods. The usual explanation given for this phenomenon is the differences in the income elasticity of demand for industrial and primary products: Engel's law. Prebisch (1950, 1959, 1963), for example, argues that the South typically exports primary products and the North industrial products: Engel's law implies a lower income elasticity of demand for primary products; hence there is a secular decline in the Southern terms of trade.[10]

The standard way to incorporate this phenomenon in a formal model is to have non-homothetic tastes. Neoclassical models of the standard trade-theory literature use this assumption to find conditions for deterioration of the terms of trade. North-South models considering capital accumulation, however, have typically not incorporated this phenomenon. Taylor (1981, 1983) lets a linear expenditure system produce non-homotheticity in the short run, but in the long-run analysis he reverts back to unit income elasticities. It would seem difficult to examine long-run properties of such models allowing for Engel's law.[11]

Engel's law as an explanation of changes in demand composition, however, may be questionable. While Engel's law is well established as a description of consumer behaviour, it is by no means obvious that Southern exports may typically be identified as primary goods with low income elasticity and Northern exports as industrial goods. Many Southern economies are large importers of food, and exporters of manufactured consumer goods; thus the terms of trade between primary products and industrial goods should not be identified with the terms of trade between the North and the South.[12] Secondly, Engel's law applies to food and not to raw materials; thus it is an incomplete explanation at best. Thirdly, the explanation completely ignores supply considerations, which suggest that there may be good reasons why the terms of trade for primary products may improve over time.[13] Torrens, Ricardo, and Malthus thought this might occur due to diminishing returns in primary production; modern versions can make some identification of primary products with exhaustible resources.

In what follows we shall focus our attention on another set of explanations of the shift in demand composition, by following Singer's (1975) lead and emphasizing differences more in the characteristics of *countries* than of *commodities*.[14] We shall simply assume that, as the relative gap between the North and the South increases, there is a shift in Northern tastes away from Southern goods towards Northern goods, and we will call it the Prebisch-Singer effect. This could occur for several different reasons, which may include the following. First, the level of development is usually associated with both improvements in the quality of products and increases in the variety of products produced; a widening in the relative levels of development will therefore result in a shift in tastes towards Northern goods.[15] Secondly, Northern consumers might *perceive* Northern goods to be relatively superior the bigger the relative gap in the levels of development.[16] Thirdly, the levels of advertising expenditures may be expected to go up with levels of development; gaps in relative development may shift tastes due to relative differences in advertising expenditure.

In terms of our model, it seems natural to measure the level of development of each country by the level of its capital stock (as reflecting the size of the industrial sector). We can therefore assume that

$$a = a(k), \tag{9.16}$$

where $a' > 0$,[17] to formalize our above arguments in our model.[18]

9.2.2 International demonstration effects

The international demonstration effect has shown up in discussions of the economic relationships between rich and poor countries, and been held as a factor responsible for uneven development. While the effect has been supposed to work in various ways, the discussions are usually vague and inconclusive.[19] We will isolate just one aspect of the argument, international demonstration effects on Southern consumption, for incorporation into our North–South model.

It has been argued by Nurkse, for example, that, due to improvements in communications the world over, Southern consumers get to know of foreign goods; this knowledge of new goods and new methods of consumption tends to raise the general propensity to consume, reduce saving, and therefore constrains Southern growth. While Nurkse emphasizes the information aspect, casual empiricism suggests also Veblen-Duesenberry type imitation effects. This implies that knowledge of foreign consumption patterns not only affects the level of consumption, but also its *composition*; this results in greater imports of Northern type goods, and creates also a balance of payments problem (which Nurkse is well aware of). Many less developed countries have tried to contain this drain in foreign reserves through the use of prohibitive tariffs; not many have succeeded due to outright smuggling and loopholes in legislation.[20]

The argument so far has focused on the nature of tastes, not changes in it. There are good reasons that may change Southern tastes over time in the course of the process of development. Such changes may be the result of both changes in the efficacy of the information distribution network, and of changes in Northern tastes. Changes in the information network reflect the greater efficacy of advertising media (TV, magazines, and the like), and may partly reflect the relative development of the North and the South. We may formalize these arguments regarding the *composition* of Southern consumption expenditure in a simple form by assuming

$$b = B(\underset{+}{a}, \underset{+}{k}, \underset{-}{q}),$$

where the signs under the argument show the signs of the partial derivatives.

This can be rewritten as

$$b = b(k, q), \tag{9.17}$$

where $b_1 = b' > 0$ and $b_2 = -1$, after substituting from (9.16).[21] q may be thought of as a shift parameter which can reduce b through trade restrictions. By a similar argument we can formalize the effect of international demonstration effects of the *level* of consumption expenditure by assuming

$$s_s = s(\underset{-\ +}{k, q}). \tag{9.18}$$

9.3 Shifts in demand composition and uneven development

We now incorporate our simple formalizations of the previous section in the basic model to examine whether and under what conditions these mechanisms can contribute towards uneven development.

The short-run properties of this modified model will be the same as those of the basic model, since, with k fixed in the short run, so are a, b, and s_s. The long-run evolution of the model, however, may well turn out to be qualitatively different.

To examine this possibility, consider the shapes of the g^n and g^s curves for this model. The g^n curve is easily seen to be the same as before, independent of the level of k. The g^s curve, however, will be quite different. To see this, substitute (9.15) through (9.18) into (9.14) and differentiate with respect to k to get

$$dg^s/dk = \{[1 + (1 - s_n)z]/(1 + z)\}$$
$$\times \{us_s(1 - a)/[s_s + (1 - s_s)b]\}(1 - e_a - e_b - e_s), \tag{9.19}$$

where

$$e_a = -d(1 - a)/dk \quad k/(1 - a),$$

the elasticity of $(1 - a)$, the share of Northern consumption expenditure on the Southern good, with respect to k, captures the strength of the Prebisch-Singer effect,

$$e_b = d\bar{b}/dk \quad k/\bar{b},$$

the elasticity of $\bar{b} = s_s + (1 - s_s)b$, the share of Southern capitalist income being spent, for consumption and investment needs, also with respect to k, captures the international demonstration effect as a whole, and where

$$e_s = -ds_s/dk \quad k/s_s,$$

the k elasticity of s_s, captures the saving impact of the international demonstration effect. Our arguments have shown that all these elasticities will be non-negative. If the international demonstration effect only affected the *level*, and not the composition, of consumption, both e_b and e_s will be positive;

if it only affected its *composition* and not its level, e_b will be positive but not e_s. The sign of expression (9.19) depends on the magnitudes of the elasticities, and is indeterminate in general. This implies that, unlike what was found for the basic model, the g^s curve may not be monotonically upward rising.

This has implications for the existence, uniqueness, and the stability of long-run equilibrium. Since the g^s line is not linearly upward rising it may never intersect the g^s line; hence long-run equilibrium may not exist. Since g^s is not necessarily monotonic it may intersect g^n more than once, so that the long-run equilibrium may not be unique. The stability of a long-run equilibrium depends on the sign of $d\dot{k}/dk$ evaluated at the long-run equilibrium, a negative sign being required for stability. Since $d\dot{k}/dk = -(dg^s/dk)k$ when so evaluated, stability depends on the g^s curve being upward rising at its point of intersection with g^n; since g^s can be downward sloping, this cannot be guaranteed.

While the exact shape of the g curve depends on the nature of the functions (9.16), (9.17), and (9.18), a possible, and perhaps likely, shape is shown in Figure 9.4 for illustrative purposes.[22] The g^s curve is drawn to allow equilibrium to exist. It is obvious that there must then be two equilibria: that at k^* is stable and that at k^{**} unstable. Let our initial state be characterized by $g^n > g^s$; at it, either $k > k^{**}$, or $k < k^*$. If we are initially at a point like k_1, development will

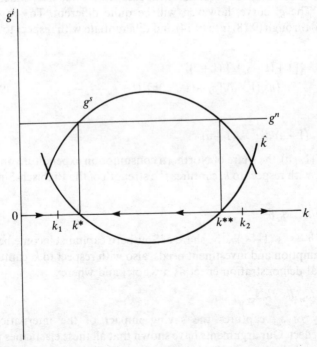

Figure 9.4   The long run in the model with endogenous preferences

be even; but if we are at a point like k_2 uneven development will occur.

Several general remarks can be made regarding the model incorporating Northern demand shifts and international demonstration effects.

First, the possibility of uneven development, which is the same as that of the instability of a long-run equilibrium, depends on the strength of the Northern demand composition shift effect and the internatonal demonstration effect, as is easily verified from (9.19). If these effects are absent and $e_a = e_b = e_s = 0$, we are back to our basic model in which development is necessarily even. Only if $e_a + e_b + e_s > 1$, at the long-run equilibrium, will uneven development occur. This happens because with the North growing faster than the South, the Prebisch-Singer effect causes tastes in the North to shift away from the Southern good, reducing Southern exports in relative terms; the international demonstration effect raises Southern imports and reduces savings, relatively, and all of this slows down the Southern rate of growth compared to the Northern rate.

Secondly, if the g^n and g^s curves are as shown in Figure 9.4, the greater the gap between the North and the South, that is, if k is higher, the more likely is uneven development.

Thirdly, there is a relationship between uneven development and the terms of trade. From (9.14) it follows that, if s_s is constant, there is a monotonic relationship between the terms of trade and g^s. This implies that, in this case, when uneven (even) development occurs, g^s falls (rises) over time, so that the Southern terms of trade deteriorates (improves) over time. This implies that, if the international demonstration effect has no effect on the saving rate, the deterioration of the Southern terms of trade is necessary and sufficient for uneven development. If the international demonstration effect also affects the Southern saving rate, then

$$dg^s/dk = (1 - V_s a_0^s) a_1^s [s_s d\pi/dk + \pi d s_s/dk]$$

which shows that it is possible for uneven development to occur even when the Southern terms of trade improves. Thus terms of trade deterioration is sufficient, but not necessary for uneven development in our model.[23]

Fourthly, if g^s and g^n do not intersect, it is possible for them to show an initial tendency to come closer together (if we start with a k at which the g^s curve is upward rising). Such a tendency should not be taken as a sign of the evenness of development, since the gap between the two rates must eventually widen.

Finally, our model has some implication for the effectiveness of government restrictions on luxury imports. Southern governments have often imposed trade restrictions on the imports of 'luxury' goods produced in the North. In our model such a shift is denoted by changes in q which affect b and s_s.[24] For a given k, such a change will have no effect on g^n, but its effect on g^s is given by

$$dg^s/dq = \{1/(1+z)[s_s + (1-s_s)b]^2\}$$
$$\times [1 + (1-s_n)z]k(1-a)u[b\delta s/\delta q + s_s(1-s_s)]$$

which implies that the g^s curve is pushed upwards, as shown in Figure 9.5. For examining the long-run effects, we should distinguish between the stable and unstable cases, as done in (a) and (b) in the figure, respectively. In the former case the long-run equilibrium value of k falls from k^* to k^{**}, so that the gap between the North and the South narrows in long-run equilibrium. If we are not initially at k^*, but at a state with $g^n > g^s$, then if $k < k^{**}$ the North will continue growing faster than the South, but at a relatively slower rate than

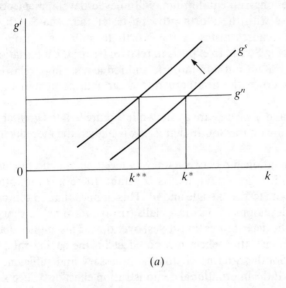

(a)

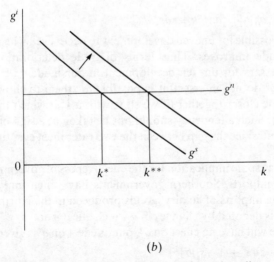

(b)

Figure 9.5   Effects of a change in trade policy

before, while if $k > k^{**}$ the South will in fact begin to grow faster and the direction of movement in k will change. In the unstable case, the change may, if $k > k^{**}$ initially, just delay in time the inevitable process of uneven development by allowing the South to 'catch up' for some time; with $k^* < k < k^{**}$ initially, the process of uneven development will be reversed to favour the South.[25] Whether k is less than, greater than, or equal to k^{**} depends, of course, on how far g^s is pushed. It is possible that even with $b = 0$ (a prohibitive restriction) $k^{**} < k$, so that uneven development can never be reversed with this kind of a policy.[26]

9.4 Conclusion

This chapter has examined a model of balanced trade between two economies, a rich North and a poor South, with differentiated structures. Defining uneven development to be a sustained increase in the ratio of capital stock in the North to that in the South, we have examined the possible role of the Prebisch-Singer effect and the international demonstration effect in resulting in uneven international development. The central results of our analysis can be summarized as follows.

1 Uneven development can occur due to the operation of the Prebisch-Singer effect and the international demonstration effect, though their existence does not necessarily imply it. The condition under which they can result in uneven development has been developed in terms of some elasticities which can in principle be measured to answer empirically the question whether they are strong enough to result in it.

2 Abstracting from all other causes of uneven development, the operation of the Prebisch-Singer and international demonstration effects will usually imply a correspondence between terms of trade deterioration for the South and uneven development. But showing that terms deterioration has not occurred cannot rule out the possibility of uneven development.

3 The South can try to stem the forces of uneven development by restricting luxury imports to some extent, but this will not work if such imports are already at very low levels. Its governments may then have little choice but to change the structure of the model, a crucial feature of which is the dependence of the South on Northern investment goods. The importance of developing a domestic investment goods sector is obvious.

In drawing these conclusions, however, we should remind ourselves of the rather special and restrictive nature of our formalization. Some of the strong assumptions, including constancy of the Northern markup, the savings rates, and technological parameters, have been made specifically to abstract from other mechanisms of uneven development.[27] Others, such as fixed coefficients, and complete specialization, have been made for simplicity. The particular model used – with the Kalecki-Steindl North and the neo-Marxian

South – does not seem crucial for our results; the main ones could have been generated using other models as well.[28] Our analysis, however, also assumed away capital mobility; further work will indicate how our conclusions may be changed by allowing for capital flows.

Apart from these implications of the model developed here, the analysis of this chapter has two methodological implications for standard neoclassical theory which takes preferences to be given. First, the model implies that, by taking preferences to be given, neoclassical economics introduces a bias towards stability and even development.[29] It is because preferences were made endogenous that our model could produce uneven development. Secondly, with endogenous preferences, it is not possible to evaluate development patterns in terms of individual or societal preferences. In general, the concept of Pareto optimality loses its sanctity.[30] We thus find that to evaluate development patterns we have to turn to 'objective' indicators such as productive capital, since evaluation by the changing 'subjective' criterion of utility is not possible any more.

CHAPTER 10

Technological change and uneven development

10.0 Introduction

While the previous chapter examined mechanisms which involve the endogenization of preferences, this chapter will examine some which require the endogenization of the other usual 'given' of neoclassical economics, that is, technology.

We will tell three stories using three models which are simple modifications of the models developed in chapter 8, one of them in fact identical to the model of the previous chapter. Section 10.1 examines a simple variant of the neo-Marxian model of section 8.2.2 and introduces technological change using Kaldor's technical progress function introduced in section 5.3.2, and shows how uneven development can arise in the sense of widening wage differentials. Section 10.2 develops a model which modifies the general framework of chapter 8 to allow for incomplete specialization, makes the North and South have identical structures – a hybrid with neoclassical and neo-Marxian features – and introduces technological change in a manner similar to Arrow's learning approach discussed in section 5.1.3. It shows how uneven development with divergent rates of accumulation can occur due to different patterns of specialization in the two regions. Finally, section 10.3 reexamines the model with a Kalecki-Steindl North and a neo-Marxian South to show how technological change due to international competition and transfer of product innovation – which could not be discussed in a closed-economy setting – may result in uneven development.

The simple models to be discussed are specific and primarily illustrative, but the stories they tell regarding technological change and uneven development are more general.

10.1 Technological change, the terms of trade, and real wages

One of the persistent themes in the discussion of North-South interaction is the asymmetric effects of technological change on the two regions. Technological change in the North serves to increase the real wage, while in the South it tends to leave Southern workers unaffected, with the benefits passed to Northern workers in the form of a deterioration of the Southern terms of trade.

As already noted in chapter 8, this deterioration has also been a persistent theme in the literature.

In this section we use a modification of the model with a neo-Marxian North and a neo-Marxian South introduced in section 8.2.2 to provide a simple formalization of this story. As a by-product we examine the dynamics behind the long-run equilibrium of that model. In the earlier presentation we did not introduce technological change into the model (except as a parametric shift). Here we have to choose a method of introducing it into the model, and we do so with Kaldor's technical progress functions, one for each region. However, we shall proceed by initially ignoring technological change, introducing it only at a later stage, after examining the dynamics and long-run equilibrium.

By the assumptions of the neo-Marxian model we have

$$1 = (W_n/P_n)a_0^n + r_n a_1^n,$$ (10.1)

$$1 = V_s a_0^s + r_s (P_n/P_s)a_1^s,$$ (10.2)

$$s_n r_n = g^n,$$ (10.3)

$$s_s r_s = q_s,$$ (10.4)

$$(1-a)[(W_n/P_n)(a_0^n/a_1^n) + (1-s_n)r_n]k = g^s + (1-s_s)br_s$$ (10.5)

which are the same as (8.12), (8.13), (8.16), (8.17), and (8.15), adding, where necessary, the neo-Marxian assumptions of full capacity utilization in both regions and fixed real wage in the South, that is, (8.21), (8.22), and (8.24).

We have not yet made an assumption about income distribution in the North. In the version of the model presented in chapter 8 we assumed that the real wage in the North (in terms of the true cost of living index) was fixed at 'subsistence'. This assumption is one that many Marxian scholars have serious problems with, especially in the presence of technological change. A more appropriate assumption could be that the labour *share* in output is fixed, and that this share depends on the state of the struggle between classes, and measures a constant rate of exploitaton.[1] This is precisely the assumption that we make here, and it implies that

$$(W_n/P_n)a_0^n = \Sigma_n,$$ (10.6)

where Σ_n, the labour share in income, is a constant.

In the short run, the stocks of capital in the two region are taken to be given, so that k is given. The levels of output are given from the relations $X_i = K_i/a_1^i$. Equation (10.6) determines W_n/P_n, (10.1) determines r_n and (10.3) determines g^n. Substitution of (10.1) through (10.4) and (10.6) into (10.5) implies that

$$P_n/P_s = \{(1 - V_s a_0^s)[s_s + (1-s_s)b]a_1^n/[(1-a)(1-s_n(1-\Sigma_n))a_1^s]\}(1/k)$$ (10.7)

which solves for the short-run equilibrium terms of trade for given k. Once P_n/P_s is determined, (10.2) solves for r_s and (10.4) for g^s.

To examine the dynamics of the economy over time, still with given technological parameters, we write

$$\hat{k} = g^n - g^s \tag{10.8}$$

which implies, upon substitution from (10.1) through (10.4) and (10.8),

$$\hat{k} = s_n(1 - \Sigma_n)/a_1^n - \{s_s(1 - a)(1 - s_n(1 - \Sigma_n))/[s_s + (1 - s_s)b]a_1^n\}k. \tag{10.9}$$

The dynamics of the economy can be shown using Figure 9.3 again, confirming the stability of equilibrium in the long run. From (10.9), the long-run equilibrium value of k is seen to be given by

$$k^* = s_n(1 - \Sigma_n)[s_s + (1 - s_s)b]/(1 - a)[1 - s_n(1 - \Sigma_n)]. \tag{10.10}$$

Substitution of this into (10.7) gives the long-run equilibrium value of the terms of trade,

$$(P_n/P_s)^* = [(1 - V_s a_0^s)a_1^n]/[s_n(1 - \Sigma_n)a_1^s]. \tag{10.11}$$

Having examined the long-run equilibrium position of the model, we now examine the effects of technological change, confining attention to long-run equilibria. We assume that each region has a linear technical progress function given by

$$(\dot{K}_i/L_i) = T_0^i + T_1^i(K_i/L_i) \tag{10.12}$$

where the indices i denote the two regions.[2] We have seen in chapter 5 that this function, together with Kaldor's 'stylized fact' that capital-output ratios are constant, implies that

$$-\hat{a}_0^i = T_0^i/(1 - T_1^i)$$

which is a constant which we will denote by h_i.[3]

From (10.10) we see that the long-run equilibrium value of k is unaffected by changes in the values of a_0^i, the only parameters of the model assumed to change due to technological progress. Thus technological progress has no effect on the nature of relative capital accumulation in our model. From (10.11) we see that

$$(\dot{P}_n/P_s) = [V_s a_0^s/(1 - V_s a_0^s)]h_s \tag{10.13}$$

which shows that the rate of growth of the Northern terms of trade is greater the higher is the rate of technological change in the South.

To examine the consequences of this technological change on real wages, note that by assumption V_s, the Southern real wage, is constant. The Northern real wage, as before, can be measured by

$$V_n = W_n/(P_n^a P_s^{1-a})$$

which implies that

$$\hat{V}_n = (W_n/P_n) + (1-a)(P_n/P_s)$$

which, substituting from (10.13), and using (10.6) implies

$$\hat{V}_n = h_n + (1-a)[V_s a_0^s/(1 - V_s a_0^s)]h_s \tag{10.14}$$

which shows that faster technological change in *both* the North and the South increases the rate of growth of the real wage in the North.[4] Thus we conclude that technological change everywhere serves to increase the ratio of the Northern real wage to the Southern real wage, and in this sense causes uneven development.

Notice, however, that in this model technological change has no effect on k^*, so that there is no uneven development in the sense of a secular increase in k over time. This particular property of the model follows from the assumption made regarding income distribution in the North. A different assumption, such as the fixed real wage assumption made for the model in chapter 8 shows that the long-run equilibrium value of k could be affected by changes in a_0^i (see Table 8.2). Rather than pursue the implications of these different assumptions for uneven development in this sense, we examine its possibility in a different model in which technological change is introduced in a different way.

10.2 Generalized learning processes and uneven development

This section examines a different mechanism by which technological change can cause uneven development in a different sense, using a model which modifies the general framework of chapter 8.

10.2.1 Generalized learning processes

It has often been argued that rich regions typically export goods involving a high degree of processing and requiring sophisticated production techniques and organization. Experience gained in the production of these goods results in the accumulation of labour skills and human capital,[5] and improvements in techniques and organization methods, the effects of which do not have to be confined to the sectors in which they originate, but may have spin-off effects on other sectors. While rich regions develops, pulled by their leading export sectors, the poor regions, specializing in the export of goods which do not require such high degrees of processing, are denied access to such vital learning processes.[6]

Some elements of this argument are found in the early writings of Hamilton (1791) and List (1841) who argued in terms of the industrial and agricultural sectors in their discussions of infant industry protection, but as explanations of uneven development it finds clear statements in Singer (1950) and Galtung (1971), for example. Singer highlights the difference between manufacturing industry and agriculture, and writes:

> The most important contribution of an industry is not its immediate product (as perforce assumed by economists and statisticians) and not even its effect on other industries and immediate social benefits (thus far economists have been led to go by Marshall and Pigou) but perhaps beyond that its effect on the general level of education, skill, way of life, inventiveness, habits, store of technology, creation of new demand, etc. And this is perhaps precisely the reason why manufacturing industries are so universally desired by under-developed countries: they provide the growing points for increased technical knowledge, urban education and the dynamism and resilience that goes with urban civilization, as well as the direct Marshallian external economies. No doubt under different circumstances commerce, farming and plantation agriculture have proved capable of being such growing points, but manufacturing industry is unmatched in our present age. By specializing on exports of food and raw materials and thus making the underdeveloped countries further contribute to the creation of industry in the already industrialized countries, foreign trade ... may have spread present static benefit fairly over both. They may have had very different effects if we think from the point of view, not of static comparative advantages, but the flow of history of a country.

Galtung talks of similar issues, although he focuses directly on the distinction between high processing and low processing industries, which may be different from the agriculture-industry distinction. He writes:

> [Consider] ... nations exchanging oil for tractors. The basic point is that this involves different levels of processing, where we define 'processing' as an activity imposing Culture on Nature. In the case of crude oil the product is (almost) nature; in the case of tractors it would be wrong to say that it is a case of pure Culture ... A transistor radio, an integrated circuit, these would be better examples because Nature has been brought down to a minimum ... The major point now is the *gap in processing level* between oil and tractors and the differential effect this gap will have on the two nations. In one nation the oil deposit may be at the water-front, and all that is needed is a derrick and some simple mooring facilities to pump the oil straight into a ship ... In the other nation the effects may be extremely far-reaching due to the complexity of the product ... It is possible to set up international interaction in such a way that the positive intra-actor effects are practically nil in the raw material delivering nation and extremely far reaching in the processing nation.

We may call this kind of trading between rich and poor countries 'vertical trading' to emphasize the hierarchical ordering of commodities by degrees of processing involved, and denote these productivity raising phenomena to be due to 'generalized learning effects'.

10.2.2 The model

To understand the implication of this phenomenon we examine a modification of our general North-South framework. There are two changes that we introduce here. First, we allow for incomplete specialization, that is, both the North and the South produce the two goods. We can thus no longer have a Northern good and a Southern good, but have a good which requires a high degree of processing (which we shall for concreteness call the manufactured good) and on involving a low degree of processing (the agricultural good). The manufactured good is both a consumption and an investment good (like our Northern good above), and the agricultural good a pure consumption good (like our Southern good). Secondly, we assume that the two regions, except for differences in the stocks of capital accumulated (and, as we shall see, consequent differences in technology), are identical. This involves assuming identical saving and spending parameters for the two regions, and allowing Southern workers (like their Northern counterparts) to consume both goods.

The first change is introduced because incomplete specialization appears an essential aspect of the story told above; spin-off effects require more than one producing sector in each region. This change also has the virtues of showing that complete specialization is not necessary for uneven development, and for allowing us to compare developments with and without trade, which was not possible in models of complete specialization in which the South could not produce without the investment good. This modification has the potential of making the analysis exceedingly complicated by having us deal with four stocks of capital (one in each of two sectors in the two regions), but we side-step this problem by assuming that the agricultural sector uses no capital in production, only labour. The second change is introduced to abstract from any other differences which could generate asymmetries in the behaviour of the two regions. This will allow us to focus on – and isolate the role of – the mechanism with which we are concerned here. The assumption of symmetry also greatly simplifies the algebra, without affecting the qualitative properties of the model.

Within this framework we will examine the dynamics of uneven development due to generalized learning effects. To do so we will need to make assumptions regarding closure, and regarding the exact nature of the learning effects.

Concerning the former, to continue to maintain symmetry we will assume that the same closure applies to both regions. This closure is not the same as any we considered in chapter 8, but a hybrid of neoclassical and neo-Marxian elements. Following neoclassical assumptions, we will assume that all workers in a particular region are fully employed, in one of its two sectors. However, we depart from neoclassical assumptions and introduce neo-Marxian elements by assuming that in the manufacturing sector the wage as given at a point in time by average earnings in the subsistence agricultural sector. This follows Lewis (1954), and presupposes that the agricultural sector is non-capitalistically organized, with workers receiving the average product.[7]

Concerning technological change due to generalized learning, we use a simple modification of Arrow's (1962) learning by doing approach. The labour-productivity ratios in the two sectors are assumed to depend on 'cumulative experience' gained only in the production of the manufactured (henceforth m) good. Experience in the production of the agricultural (or a) good has no such effect. In section 5.1.3 we stated that in learning by doing models such experience can be measured either by cumulative output or cumulative investment. As in the earlier chapter we follow Arrow by considering cumulative investment, or what is the same thing in the absence of depreciation, capital in the m sector.[8] We make a further modification of scaling capital by the total labour force, to take into account the possible 'thinning-out effect' of a higher population. Assuming log-linear forms as before, we have

$$b_{mj} = \Theta_m k_j^{n_m}, \tag{10.15}$$

$$b_{aj} = \Theta_a k_j^{n_a}, \tag{10.16}$$

where $n_i < 1$ to allow for diminishing returns to learning and Θ_i and n_i are positive parameters of the generalized learning function, and where j refers to the region, $j = n$ or s.[9] The output-capital ratios are assumed to be constants as before. Note that we have incorporated the assumption that the parameters of the learning function are identical between the regions.

We now examine the structure of the model.[10] We have two regions, a North and a South, which we denote by n and s, and two goods, a manufactured good and an agricultural good, which we denote by m and a. The a good requires only labour in production, while the m good requires both capital and labour, in fixed proportions; both use a constant returns to scale technology. The total supplies of labour in each region are fixed at a point in time, while the stocks of capital are fixed at a point in time as a result of past investment. Perfect mobility of labour within each region is assumed; the real wages in the two sectors are therefore equalized in equilibrium as long as there is incomplete specialization.

For each region j, $j = n, s$, we therefore have

$$X_{mj} = cK_j, \tag{10.17}$$

$$L_{mj} = X_{mj}/b_{mj} = cK_j/b_{mj}, \tag{10.18}$$

$$L_{aj} = L - L_{mj} = L - cK_j/b_{mj}, \tag{10.19}$$

$$X_{aj} = b_{aj}L_{aj} = b_{aj}(L - cK_j/b_{mj}), \tag{10.20}$$

where X_{mj} and X_{aj} are the outputs of the m and a goods, c is the output-capital ratio in the m sector (the a sector uses no capital) assumed to be the same across regions, b_{ij} the productivity of labour in sector i,[11] L_{ij}, the employment of labour in sector i, K_j the stock of capital in sector m, and L the total available labour force (assumed to be of the same size in each region).

Let the price of the m good in terms of the a good be P_m. We shall assume throughout that neither region is completely specialized in production. With this assumption, in each region the wage in terms of the a good is equal to the average product of labour in that sector. With the m good being the capital good, the rate of profit in the m sector is given as

$$r_j = (P_{mj}X_{mj} - b_{aj}L_{mj})/P_m K_j = (1 - b_{aj}/P_m b_{mj})c. \tag{10.21}$$

Given the assumption that all wages are consumed and that a fraction s of profits (in both regions) is saved and automatically invested in the m sector in which it is earned, we have

$$\hat{K}_j = sr_j. \tag{10.22}$$

We assume that the labour force in each region grows at the rate of growth n, implying that

$$\hat{k}_j = (sr_j - n). \tag{10.23}$$

We now examine the behaviour of the international economy portrayed by this model, first under autarky, and then under conditions of free trade.

10.2.3 International development under autarky

Under autarky, short-run equilibrium – which clears the commodity and labour markets for their given stocks of capital – implies (in what follows each equation will be for $j = n, s$)

$$P_{mj}X_{mj} = ab_{aj}L + [s + (1-s)a]r_j P_{mj}K_j. \tag{10.24}$$

This implies the clearing of the m goods market, Walras's law taking care of the a goods market. Substituting from (10.17) to (10.21) we get

$$P_{mj} = [ab_{aj} - \tau(b_{aj}/b_{mj})ak_j]/(1-\tau)ck_j, \tag{10.25}$$

where $\tau = s + (1-s)a$. In the short run k_j is given, so that (10.15) and (10.16) determine the b_{ij}, so that P_{mj} is determined. The values of X_{mj} and X_{aj} can then be found from (10.17) and (10.21).

From (10.15), (10.16), (10.21), (10.23), and (10.25) we can see that, over time, k_j changes according to

$$\hat{k}_j = s\{1 - (1-\tau)c/[a\Theta_m k_j^{n_m - 1} - \tau c]\}c - n. \tag{10.26}$$

Long-run equilibrium will be attained when the right-hand side of the above equation vanishes. This equilibrium can be shown to be stable when $n_m < 1$. This implies that, starting from any initial k_j, the two economies will move over time, isolated from each other, and attain a long-run rate of growth n, with an identical k_j given by

$$k_j^* = \{a\Theta_m(cs - n)/c(cs - \tau n)\}^{1/(1 - n_m)}. \tag{10.27}$$

We assume that the existence conditions are satisfied, or that $sa > n$.[12]

10.2.4 *International trade and uneven development*

Having examined the behaviour of the international economy under autarky we turn to its behaviour when the two regions trade with each other. With free and balanced trade, equilibrium implies

$$P_m(X_{mn} + X_{ms}) = a(b_{an}L_n + b_{as}L_s) + \tau[P_m - (b_{an}/b_{mn})]cK_n$$
$$+ \tau[P_m - (b_{as}/b_{ms})]cK_s$$

which implies that

$$P_m = \{a(b_{an} + b_{as}) - \tau c[(b_{an}/b_{mn})k_n$$
$$+ (b_{as}/b_{ms})k_s]\}/c(1-\tau)(k_n + k_s). \tag{10.28}$$

Substitution from (10.15) and (10.16) for given k_j will solve for the short-run equilibrium terms of trade which clear all markets.

We can then completely determine free trade equilibrium given K_n and K_s; the levels of production, export, import, and consumption in each region can be solved.[13] The direction of trade can be found from comparison of the autarkic prices given by (10.25). Under reasonable parameter restrictions it can be seen that $k_n > k_s$ implies $P_{mn} < P_{ms}$, so that the North will export the m good and import the a good.

From (10.15), (10.16), (10.21), (10.23), and (10.28) it follows that the dynamics of k_j in the long run are given by

$$\hat{k}_j = s(1 - Q_i)c - n, \tag{10.29}$$

where

$$Q_i = c(1-\tau)(k_n + k_s)/[a\Theta_m(k_i^{nm} + k_i^{nm-n_a}k_j^{n_a})$$
$$- \tau c(k_i + k_i^{nm-n_a}k_j^{1-nm-n_a})]$$

for $i = n, s$ and $i \neq j$.

At long-run equilibrium we have the right-hand side of (10.29) equal to zero, which implies that

$$k^* = k_n = k_s = \{a\Theta_m(cs - n)/c(cs - \tau n)\}^{1/(1-n_m)}. \tag{10.30}$$

It can be seen that the long-run equilibrium with free trade is the same as under autarky. But will the two economies actually reach this long-run equilibrium when they trade, starting from some initial stocks of capital?

To address this question we examine the stability of this long-run equilibrium for the system under free trade by examining the properties of the Jacobian of the system in the neighbourhood of the equilibrium (we are therefore considering local stability properties only).

The Jacobian of this system given by (10.29) is seen to be given by

$$\begin{bmatrix} s(1-Q_n)c - n - sck_n(\delta Q_n/\delta k_n) & -sck_n(\delta Q_n/\delta k_s) \\ -sck_s(\delta Q_s/\delta k_n) & s(1-Q_s)c - n - sck_s(\delta Q_s/\delta k_s) \end{bmatrix}.$$

At long-run equilibrium its trace is given by

$$-[sc^3 2k^{*2}(1-\tau)/(cs-n)D^2][cs(1-2n_m+n_a)+\tau cs(n_m-n_a)$$
$$-\tau n(1-n_m)], \tag{10.31}$$

where D is the denominator of Q (since the k_j are equal, we write $Q_j = Q$), and its determinant given by

$$-[s^3 c^6 16 k^{*3+nm}(1-\tau)^3 a\Theta_m/D^4](1-n_m)(n_m-n_a)/(cs-n). \tag{10.32}$$

To fix ideas, we first consider the case in which capitalists – in both regions – consume very little of the agricultural product, so that a is near unity, implying that τ is close to unity as well.

In this case it is easy to see that a sufficient condition for the long-run equilibrium being a saddle-point is that $n_m > n_a$. In this case, since $n_m < 1$ (due to diminishing returns to learning) and $cs > n$ (as assumed above), the determinant given by (10.32) is negative. For $\tau - > 1$ the trace can be seen to be negative as long as $n_m < 1$. It follows that the long-run equilibrium is a saddle-point as shown in Figure 10.1. The phase diagram in that figure shows that the $dk_n/dt = 0$ and $dk_s/dt = 0$ lines are both downward-sloping, and that they intersect as shown. Unless the international economy happens to start from a point on the separatrix SS, it will find itself on a path on which (ultimately) one of the k_i increase indefinitely over time, the other falling. If we start from above the separatrix – even a hair above it – so that $k_n > k_s$, we will end up ultimately with k_n rising over time and k_s falling. Since L rises in both

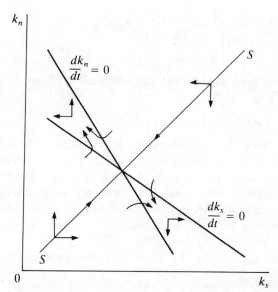

Figure 10.1   Uneven development with generalized learning

regions at the rate n, this implies that K_n/K_s will increase over time. There will thus be uneven development in this sense. The initial advantage to the North, however small, which is manifested in $k_n > k_s$ (note that this is the *only* difference assumed between the two regions), cumulates over time. This occurs due to the fact that the North's advantage in having a higher capital-labour ratio makes it have a higher productivity in both sectors, but this higher productivity makes it have even greater production experience in the manufacturing sector, which, due to the generalized learning effects, puts it further ahead.

We proceed with the examination of two other possible cases, to explore the limits of the applicability of this special case. In the first, assume that τ is considerably smaller than unity. In this case, (10.31) shows that the trace may be positive if n is sufficiently small and if $n_m > (1/2) + (n_a/2)$. In this case the phase diagram is shown in Figure 10.2(a), which is once again seen to be a saddle-point, implying that this case shares the qualitative properties of the previous one.

In the final case, we assume $n_a > n_m$. In this case (10.32) shows that the determinant is positive. Further since with $n_a > n_m$ it is impossible to have $n_m > (1/2) + (n_a/2)$ or $1 - 2n_m + n_a < 0$, it follows from (10.31) that the trace is negative. Thus the stability conditions are satisfied, and the phase diagram is as shown in Figure 10.2(b). This implies that $n_m > n_a$ is a necessary condition for the absence of stability in the model.

We may conclude from this examination that the condition for uneven development to occur is $n_m > n_a$, or that the elasticity of the productivity of labour with respect to capital in manufacturing per worker in the economy is greater in the manufacturing sector than in the agricultural sector. In this case if the North starts off with even a slightly higher capital-worker ratio, it will eventually experience a growing capital-labour ratio, with the South experiencing a declining one. Since the capital-worker ratio, according to (10.15) and (10.16), affects the level of technology in each sector, and technological change and capital accumulation are the sources of growth in this model (outside long-run equilibrium), we can conclude that the uneven development which occurs hurts the South and helps the North. If the uneven development condition is not satisfied, and $n_a > n_m$, then with greater capital accumulation by the North, the productivity in agriculture rises faster than in manufacturing. This rise in productivity in agriculture, by our assumptions, increases the wage rate in manufacturing, and by squeezing profits, serves to introduce an equalizing force to the accumulation process. In this, the international economy eventually reaches the long-run equilibrium at which both regions reach the same capital-worker ratio and grow at the rate n.

In addition to the fact that our model shows the possibility of uneven development in the presence of endogenous technological change, it shows that the South (if it starts with a lower capital-worker ratio and the uneven development condition is satisfied) can actually lose from trade. It will do

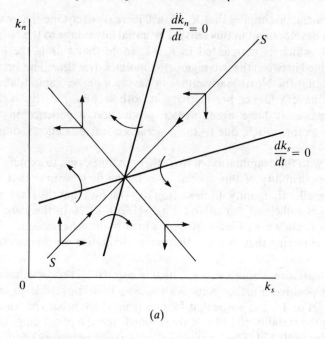

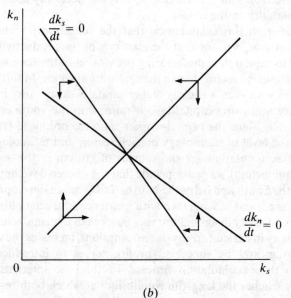

Figure 10.2   Additional cases of international development with generalized
learning

better by closing itself from trade and thereby shunt itself from an uneven development path where its capital-worker ratio continuously falls over time, to a stable path where the capital-worker ratio eventually stabilizes at its long-run equilibrium value.

Because of the rather striking nature of this conclusion, we end this section with some remarks qualifying it.

First, the South can gain in this model by becoming autarkic in the sense that it will break out of the process of uneven development. This does not imply that it will do be better in the standard trade-theory sense, or that it is hurt by trade in that sense. In evaluating gains and losses from trade in the international trade theoretic tradition, one usually considers utility from consumption bundles, or in dynamic analyses, present discounted value of consumption streams. Our approach has focused only on the capital-worker ratio (or stock of capital) in each region, and immiserization has been defined as a reduction in this stock of capital. If we are to argue that this kind of immiserization implies that regions lose from trade, we must either argue that it is better to look at welfare in terms of capital than in terms of utility from consumption bundles, or that losses in terms of one are associated with losses in terms of the other.

One can make the argument that the capital-worker ratio may be a better measure of the gains and losses. If we consider the reasons for the growth of the productive potential of our economies we find that they have to do with the accumulation of capital and technological change, and that both of these are intimately connected with the capital-worker ratio in each region. If we accept this argument about the 'objective' indicator of welfare being superior to the 'subjective' indicator based on utility, we can accept that losses from trade can occur. The arguments of the previous chapter, based on the endogeneity of preferences, strengthens the argument being made here.

If we do not accept this argument, we have to compare utility streams with and without trade, which is extremely difficult, since we would have to consider the familiar initial gains from new price ratios (which are not considered in this model), and changes in the terms of trade over time, in addition to the effects of capital accumulation and technological change. However, it is easy to see that, even if we have an unstable configuration, it is possible that, with the deindustrialization of the South and the eventual specialization of the North in the production of the manufactured good, the terms of trade can move strongly in favour of the South. The real income and hence consumption in the South may thus not necessarily fall in the South in this case, even though its productive potential is diminished. However, a loss in the utility sense clearly becomes possible for the South, remembering the fact that the manufactured good is the investment good, the demand for which rises with the development of the North, and if in addition we allow the demand for the agricultural good to be highly price elastic.

Secondly, the demonstration that the South can actually lose from trade does not really contradict the gains from trade theorems. The latter state that a region will gain from trade in the absence of any 'distortions'; if there are 'distortions' the region will gain if it follows 'optimal' policies. However, economists who refer to them usually believe that economies do gain from trade: either the distortions are judged to be empirically unimportant, or they can be 'corrected' by the use of optimal policies.[14] The existence of generalized learning effects on our model implies some sort of a dynamic production externality (or can be viewed as giving an economic rationale for a 'non-economic' production objective), so that it should really be of no surprise that in the absence of optimal production taxes/subsidies losses from trade can occur. From this point of view, our analysis serves to show that such production externalities are not pathological curios but important elements of the development process, and that given such distortions, immiserizing trade will occur under certain identifiable conditions.

Finally, it is of interest to emphasize that the South loses without being the victim of overt colonialistic exploitation by the North, or even of imperialistic exploitation in the Galtung (1971) sense with the 'centre' in the South serving as the tool by which the 'centre' in the North destroys the South (the North and South each having a periphery and a centre). Trade occurs and hurts merely because it is in the myopic interests of all consumers who want to buy from the cheapest seller. However, there is exploitation of the South by the North in a sense which is a generalization of Roemer's (1982) definition: it is possible for the South to withdraw from the trade relations and gain in a long-run dynamic sense.

10.3 Product innovation and uneven development

The models discussed in this chapter so far have two features in common. First, they abstract from all consideration of aggregate demand problems. Secondly, they assume that technological change in any region depends only on developments within that region. It may – with some justification – be thought that these features bias our conclusion towards uneven development. In chapter 5 we saw that in the presence of aggregate demand problems more rapid technological change involving the reduction of labour-output ratios may be immiserizing. Consequently, it is possible that the ability of the North to generate more rapid technological change may immiserize it relatively to the South, and this would actually serve to mitigate the forces of uneven development. Regarding the independence of technological change in the two regions, it may be argued that, once technology transfers are allowed for, the improved technology of the North can be transferred to the South, which again may serve as an equalizing mechanism weakening the forces of uneven development.

While there is some truth in these arguments, it does not follow that there are no tendencies towards uneven development if demand factors and technological interdependence are brought into the analysis. We have already seen in chapter 5 that, in models where demand problems exist, technological change may be growth-inducing by increasing aggregate demand. In a North–South framework it is also possible for such an effect to operate through the composition of spending. To illustrate this argument we need only reinterpret the model of the previous chapter, in which the Northern economy had a Kalecki-Steindl structure and was therefore demand constrained. Consideration of this model will also show that in some ways technological interdependence could strengthen, rather than abate, the forces of uneven development.

The reinterpretation of that model starts from a recognition of the fact that, even though we have been assuming only one Northern good and one Southern good in our model, we may think of the Northern and Southern goods as actually comprising of a large number of products, each of which are consumed by Northern consumers and Southern capitalists (our assumptions for that model allow Southern workers only to consume the Southern good). All the products are viewed by these consumers as being identical so that they enter symmetrically into their preferences (although the amounts purchased need not be the same because of different prices), and all Northern products and all Southern products are assumed to have the same price, although the Northern and Southern terms of trade can be different from unity.[15] Consumers therefore spend a fixed proportion of their total consumption expenditure on each product. The proportion spent on Northern products and Southern products will consequently depend on the number of Northern and Southern products. We assume, for simplicity, that saving rates are independent of the number of products.

The number of products produced in each region changes due to product innovation which is the result of technological change. Product innovation can occur in the North and the South as a result of research and development, processes which can occur independently in the two regions in the same way as discussed in the chapter so far. However, for this model we can also interpret such innovation as resulting from technology transfers. The South can receive technology from the North through multinationals or licensing agreements, while there can also be a reverse transfer of technology due to the so-called brain drain.

Rather than enter into the details of each of these mechanisms, we simplify by relating the ratio of the number of Northern products to Southern products, denoted by y, to the ratio of Northern capital to Southern capital, k, and explore the likely direction of this relationship with the following remarks.

1 Product innovation is largely dependent on the amount of resources that can be expended on research and development activities. If the availability of such resources is positively related to the stock of productive capital or the level of economic activity then y is positively related to k. If the availability of resources is positively related to the level of per-capita

output, as long as population growth rates are fixed, the same relation will hold.

2 Product innovation can also be said to depend on demand considerations, that is, the availability of markets. If innovators think that the important markets – at least initially – are domestic ones, there is a positive relation once again between y and k.

3 The transfer of technology from the North to the South partly depends on the gap between the North and the South. Veblen (1915) and Gerschenkron (1962) argued about the beneficial effects of the late starter; the greater the gap the more the potential of catching up, and hence the greater the transfer. While this argument may be valid for regions that are not too far apart, it seems less applicable for the North and the South, where the technological gap is, except for the more industrialized Southern region, too large. For North–South relations, an excessively large gap implies that the South will not be able to effectively receive technology from the North and develop new, exportable, products unless its own technological base is relatively well developed. If we measure the level of technological sophistication with the stock of capital, this implies that y is again directly related to k. More rapid technology transfer will reduce y, which is possible only if the technological gap between the North and South closes and k is lower.

4 The brain drain has often been seen as a reverse transfer of technology from the South to the North. Due to better economic opportunities and the availability of more satisfying and challenging jobs, Southern scientists, scholars and technicians migrate to the North, and thereby serve to widen the technological gap between the North and the South. If the availability of economic opportunities and better jobs depends on the stock of capital, the brain drain will be greater the higher the k, and consequently, there will be a higher y.

Due to the mechanisms discussed here, we find that there are reasons to expect that an increase in k will result in a higher y over time, and this will imply an increase in a and b. There may of course be countervailing tendencies at work, but we assume that the mechanisms just described are strong enough to make us assume that a and b are positively related with k over time, exactly as in the previous chapter. If these relationships are strong enough to satisfy the elasticity condition discussed in that chapter (noting that s_s is constant), it follows that product innovation can result in uneven development in the sense of an increase in k over time.

10.4 Conclusion

In this chapter we have examined the possibility of uneven development on an international scale due to technological change. To do so we have examined three models of North–South trade.

In the first, we introduced Kaldor's technical progress function into a North-South model in which both North and South were neo-Marxian, although in somewhat different senses; the South had a real wage fixed at subsistence, while the North had a fixed wage share determined by the state of class struggle. Technological change in both the North and the South raises the real wage in the North and results in a secular decline in the South–North terms of trade. Uneven development occurs in the sense of a widening real wage gap between the two regions.

In the second, we introduced a modification of Arrow's learning function into a model of incomplete specialization which resulted from only a slight modification of the framework developed in chapter 8. We saw that uneven development in the sense of divergent rates of capital accumulation is a possibility if technological learning (including spin-off effects) occurs primarily in one sector of the economy, the manufacturing sector. The international pattern of specialization can result in a situation in which the North actually accumulates more and more capital per head while the South experiences a decline in capital per worker. While this denouement was seen to be a distinct possibility, it is not inevitable. Whether or not uneven development occurs depends on the parameters of the learning functions.

Finally, we found that uneven development due to technological innovation was possible also in a model in which the North was Kalecki-Steindlian and the South neo-Marxian, exactly as in the previous chapter. Our discussion reinterpreted the model of the previous chapter to discuss uneven development due to divergent rates of product innovation, introducing also the mechanisms of technology transfer and the brain drain, although in a cursory manner.

While our models have been illustrative and our analysis of particular mechanisms only too sketchy, they serve to show that the introduction of technological change into North-South models can, in a variety of settings and through a number of technological mechanisms, result in uneven development. The inequalizing effects of international trade have to be given serious attention if we are concerned with reducing the gap between the rich and poor nations of the world.

CHAPTER 11

Conclusion

In this book we have been concerned mainly with four alternative models of growth and income distribution. We started by considering these models as alternative ways of 'closing' a simple one-sector, closed-economy underdetermined framework for studying growth and distribution in a capitalist economy. We then considered the four approaches – neoclassical, neo-Marxian, neo-Keynesian, and Kalecki-Steindl – in a more general way, and related them to our models, modifying some of our simplifying assumptions. Subsequently we introduced additional complications, including monetary issues and inflation, technological change, and attention to sectoral issues, and relative prices. Finally, we extended the analysis to the study of trade between two regions to examine issues relating to uneven development on a world scale, in particular, changes in preferences and technology.

The different models had remarkably different implications. This was true for the relationship they implied between growth and distribution, for the effects of a higher rate of monetary expansion, the implications of greater technological dynamism, the role of competition in the determination of relative prices, and for the mechanics of uneven development, to mention only a few implications considered in this book.

We conclude with three remarks, two on ways in which the analysis of the book ought to develop, and one in which – at least at present – it should not.

First, the models we have been analysing in this book are extremely simple ones, analysing basic issues in a transparent manner. It would be useful to extend them to make them more rigorous. Natural extensions include introducing assets in a more careful manner, and conducting rigorous multi-sector analysis.

Secondly, we have discussed only a few factors which have a bearing on growth and distributional issues. It would be useful to extend the analysis to examine the role of factors such as population growth, land and other natural resources, and government policies.[1] The analysis of their effects on growth and distribution are usually made in terms of neoclassical models, and their analysis in terms of the other models would improve our understanding of their role.

Thirdly, we have examined alternative models which have very different implications, but not attempted to choose between them, either with rigorous

214

empirical work, or on grounds of intuitive appeal. The approach of this book is to deliberately eschew making such choices, and to examine a diversity of models and mechanisms. Our intuitions can then be broadened rather than shaped by a particular theory on which we happen to have been raised, and we can develop more adequate bases for the empirical examination of different theories instead of being slaves to empirical concepts which are so closely shaped by established theory as to make alternatives seem 'unrealistic'. We may be more attracted to one of the different approaches for a variety of reasons: because they give the answers we like to hear on ideological grounds, or because our thinking is too strongly shaped by knowledge of one theory and ignorance of the others, or because we are iconoclastic and therefore want to attack established theory. But the progress of economics requires that they are all studied and developed, and further, that economists concerned with issues relating to growth, distribution and uneven international development have a chance to have their intuitions shaped by all the different approaches.

NOTES

1 Introduction

1 We do not use any formal criteria to discover the essence of each approach. As will be clearer as we proceed, the essence of each approach will be that which can be easily identified as a special case of our general framework, and this procedure is ultimately justified empirically in a loose manner by examining some standard representations of each approach.

2 Similar criticisms have been made of Marglin's (1984a, 1984b) efforts (Nell, 1985, Hahn, 1986). I have argued elsewhere (Dutt, 1987a) that Marglin's approach can be absolved of some of the criticisms by using a different general framework from the one proposed by him, and using different closing rules.

3 Popper's own ideas, of course, are far more complex and sophisticated, showing an awareness of many of the problems mentioned here; what is caricatured here is the standard mainstream economists' view of Popper, which are rooted in some of Popper's ideas.

4 See Hicks (1979) for a brief discussion.

5 See Cross (1982) for a discussion with a macroeconomic example.

6 See Cross (1982) for problems in identifying the hard core of a research programme.

7 See, for example, Weintraub (1985).

8 Use of individual optimizing behaviour in a formal way makes this method a type of methodological individualism, but is not equivalent to assuming that individuals are rational. Using optimizing behaviour simply implies that the neoclassical economist takes a conventionally acceptable objective function and choice set and assumes that agents optimize. If the choice set does not capture the important constraints faced by the agents, or the objective function is inconsistent in some way (the use of a positive rate of time preference by individuals is a possible candidate for inconsistency), optimizing behaviour can be said to be not rational. If a non-neoclassical economist does not use optimization but instead assumes behavioural rules, this does not imply irrational behaviour. First, the behavioural rules may be derivable from an optimizing exercise. Secondly, and more importantly, given a complex environment, it may be most rational for agents to fall back on rules of thumb and satisfice: the cost of gathering all the relevant information may be excessively high to warrant such gathering for setting up an optimizing exercise. Since an agent cannot know a priori how much information to gather, they must in effect be following rules of thumb all the time. To force them to optimize in economic models implies choosing arbitrary objective functions and oversimplified constraint sets. For an illuminating discussion of these issues, see Elster (1979).

9 Several candidates can be suggested. Thus a reduction to 'psychological factors', 'nature of uncertainty in monetary economies', and 'the problem of effective

demand' in the neo-Keynesian approach, and the 'economic surplus theory' or 'the problem of effective demand' in the Kalecki-Steindl approaches can, and have been, suggested. (Emphasis on effective demand obviously blurs the distinction between these two approaches.) But these organizing principles do not have the same theoretical status or widespread acceptance within each group as the ones suggested in the text.

10 Although in many cases it would be difficult to draw such a line in a clear way, we believe that it is useful to attempt to do so.

11 It is not important for our purposes as to whether a theory has been, or can be, depicted with the use of a particular set of equations. Our use of modelling language is only for making the idea of the dichotomy precise. Issues regarding dichotomies will be considered in more detail in chapter 7.

12 It is not always obvious which one the theorist actually starts with, but this is not important for our purposes. It should be noted that not all dichotomies necessarily imply particular visions, since some dichotomies are essentially heuristic in nature (for instance, that made in neoclassical partial equilibrium analysis). However, even these types of dichotomies may imply the retention of specific visions when the theory develops, as we shall presently argue.

13 Empirical verification raises all the problems discussed earlier in discussion of Popperian method. Cross's (1982) suggestion of following some parts of Lakatosian method at best only addresses the Duhem-Quine problem mentioned above.

14 Solow (1985) continues just following the passage quoted above, with: 'I hope no one here will think that this low-key view of the nature of analytical economies is a license for loose thinking. Logical rigor is just as important in this scheme of things as it is in the more self-consciously scientific one ... The case I am trying to make concerns the scope and ambitions of economic model building, not the intellectual and technical standards of model building.'

15 Pasinetti's (1981) analysis can be said, in parts, to lack behavioural content. This does not imply that the analysis cannot be made consistent with particular behavioural hypotheses; only that these are not usually carefully explored.

16 The *locus classicus* is Taylor (1983), although this work has several important precursors. Contributors to this tradition – the present author included – do not claim that they understand completely the structure of an economy that they study, or that they can defend their particular models using econometric techniques. Their approach is usually to tell stories about what they believe are important mechanisms at work in particular contexts, and to make assumptions on the basis of 'stylized' facts. However, in so far as they often use their models to do simulation work using actual numbers, they implicitly seem to be wedded to a particular set of economic characteristics, for a particular context, revealing their preference for one type of model over another. The stance taken in this book is that, while it is all right, and perhaps inescapable, to do this, one must be aware that the grounds for choosing between different models in a given context are rather flimsy, at least given our present state of knowledge. Thus one should be aware of the implications of alternative models.

17 We are thus more concerned with the macro-foundations of microeconomic behaviour rather than with the currently popular micro-foundations of macro-economics.

18 See Dow (1980).

19 For a contrary view, which we cannot entertain, see Shapiro (1978).

20 If this sounds too strong, it is because we are using the word equilibrium in a sense

different from the way it is usually used in economics. Thus disequilibrium models in macroeconomics usually refer to models having positions of rest which are different from the Walrasian notion of market clearing; they are thus equilibrium ones by our definition. Actually, our method of confining attention to equilibrium positions is not restrictive at all, although the particular equilibrium models we use may be. The non-restrictive nature of the equilibrium approach follows from what Schlicht (1985, 45–6) calls the Hicks-d'Alembert principle which states that 'any state can be viewed as an equilibrium state by referring to suitable additional influences'.

21 The terms 'run' and 'period' will be used synonymously to denote logically more or less restrictive models, where more restrictive implies freezing more variables.

22 See Dow (1985).

23 Additional issues, such as irreversibilities, could be introduced to make some equilibria path dependent, to make the models mimic history more closely.

24 This type of causality corresponds to Hicks's (1979) notion of contemporaneous causality. We believe that this notion of causality is appropriate for discussing the determination of growth and income distribution which concern the study of dynamics. Hicks (1985) argues that models of steady states – which we will often use – are not dynamic ones, using a particular meaning of the term dynamic. We believe that we are doing dynamic analysis in this book despite the preponderance of steady state models for the following three reasons: (1) we are concerned with capital accumulation and technological change, which are dynamic phenomena; (2) we often examine the process of adjustment – through short run equilibria – to notional steady states; and (3) our steady states are notional both in the sense that the economy can hover around it and the steady states themselves can move over time.

2 Alternative models of growth and distribution

1 This follows the presentation in Dutt (1987a), although the present version allows capitalists to consume while the earlier one did not.

2 Any equilibrium solving for $g > 0$ implies that K is changing. We will later define a *short-run* equilibrium as one in which K is given. Hence this equilibrium is called a *long-run* one.

3 See Pasinetti (1962), Darity (1981) and Dutt (1989c) for analyses of the implications of saving and capital ownership by workers in some of the models considered in this book.

4 A precise mechanism will be examined when we consider the short-run analysis in the next section.

5 This is a necessary, but not sufficient, condition for full-employment growth, since it is consistent also with a constant unemployment rate.

6 The standard neoclassical result that real wage equals marginal product of labour (and a similar result for capital) is not obtained in this model since with fixed coefficients, that marginal product is zero where it is defined.

7 It is not being claimed that these assumptions do justice to Marx's own analysis. However, these, or similar assumptions (see the discussion of the next chapter) are behind neo-Marxian models examined by Marglin (1984a, 1984b), Bauer (1986), and others, although these writers often introduce additional features not examined here. The analysis here also abstracts from labour values.

8 See, for example, Shaikh (1978, 1980), and chapter 7 below.

9 A Ricardian justification to the fixed wage assumption can be given in terms of Malthusian population dynamics. Lewis (1954) defended his use of the fixed real wage assumption, in part, by pointing to the existence of disguised unemployment or surplus labour.

10 This is particularly innocuous in the long run, but we will maintain the same assumption in analysing the short run, following Keynes's (1936) view that present conditions exert a strong influence on firms in their estimation of an uncertain future.

11 We could alternatively distinguish between g^d, desired accumulation, and g, desired accumulation, denote the desired accumulation function by g^d, and have another equation $g = g^d$. Distinguishing the two will be important in chapter 4 below.

12 See Harris (1978).

13 This model follows Dutt (1984a), an earlier 1981 draft of which was incorporated in Dutt (1982). A similar model, developed by Rowthorn (1982), came to my attention after mine was developed.

14 All our results could have been obtained by removing r as an argument, and using an accelerator-type model of investment (involving output levels rather than changes). See Amadeo (1986).

15 If we had assumed, for example, that capitalist consumption was in part dependent on profit and in part dependent on the capital (wealth) held by the capitalists, and was linear, our saving function could be written as

$$S = -aK + srK$$

which would imply

$$S/K = -a + sr$$

where $a > 0$. The saving curve would now have a negative vertical intercept, so that it follows that an investment curve with a smaller slope than it and with a negative vertical intercept could intersect it in the positive quadrant.

16 It is obvious that we needed to keep V fixed in the short run to avoid it being driven to the extremes, $V = 0$ with unemployment and $r = 0$ with excess capacity, given our fixed coefficients assumption.

17 To save notation, we shall henceforth denote constant, positive, adjustment coefficients in different models by the floating symbol Ω. In models with more than one adjustment coefficient, we will distinguish them by subscripts.

18 The cycles – somewhat surprising for a neoclassical model – are generated by the non-neoclassical assumption of short-run wage rigidity.

19 If we let V change over time, so would r and g. When $V = \bar{V}$ the levels of r and g would come to their long-run equilibrium levels.

20 This analysis suggests a new closure of our framework, given by equations (2.1), (2.3), (2.4), (2.6), and (2.13). In this model n determines g. Although this is a consistent model we have not given it the pride of place alongside the other models because it has not attracted much attention in the literature. It also seems odd that capitalists should invest all their saving when they hold idle capacity: a desired accumulation function seems to cry out for recognition. The model can be rescued by assuming that the market shares of individual firms are proportional to their investment.

21 The two curves intersect at $N/K = (a_0/\tau)[n(1 - \beta/s) - \alpha]$. The case in which the two are equal is easy to deal with, as is the case in which the n curve is always below the g curve.

22 This model does not insist on full employment growth. More faithful to Johansen is

a model of full employment growth which endogenizes the real wage. The model would be the same as this one, with (2.8) replaced by (2.6); we would also need to introduce fiscal factors explicitly.

23 This follows Dutt (1986c) and abstracts from changes in inventories. While Kalecki and Kaldor did not use the Marshallian concepts of demand and supply price, their use seems to provide a general framework which synthesizes their approaches. Note that, if output changes are very rapid, the actual price need not depart at all from P_s, making our analysis identical with Kalecki's.

24 See Dutt (1986c).

3 Alternative models and alternative approaches

1 This is more than methodological individualism; it requires explicitly writing down explicit objective functions and constraints and deriving behavioural functions from the optimization of the objective functions given the constraints.

2 But many would recognize features which superficially may seem like 'distortions', but which upon closer inspection (and with imaginative and resourceful economic agents) do not seem to cause market failure. Coase's theorem, tournaments, and auctions could thus dispose of 'distortions' due to externalities and asymmetric information, for example.

3 See, for example, Arrow (1962), Stigler and Becker (1977), Elster (1979) (and references on endogenous changes in preferences cited in it), Akerlof (1984), and Radnitzky and Bernholz (1987).

4 Take Solow, one of the foremost neoclassical economists. His growth model (Solow, 1956) assumes a constant saving-income ratio not based on optimizing behaviour, and in another context (Solow, 1979) he takes as a 'primitive' the wage-efficiency relationship. In both cases, of course, others have rushed to provide microfoundations in terms of individual optimizing behaviour. It is doubtful whether an approach in which all behaviour can be reduced to the optimizing approach is feasible. One obvious example: in standard competitive general equilibrium theory, the auctioneer does not optimize.

5 See, for example, Hahn (1986).

6 See Boland (1981).

7 See Caldwell (1983).

8 See, for example, Burmeister (1980).

9 Perhaps of greater use in the consideration of models of individual actions and responses, than in macroeconomic models such as ours.

10 Although because we have differential saving behaviour aggregate 'preferences' change endogenously with income distribution.

11 The nature of modifications may be considered problematic (or, as noted in chapter 1, may be doomed to be so), but that is another matter.

12 There is no denying that neoclassical economists have made many contributions – too numerous and well known to enumerate – to the economics of imperfect competition and unemployment. However, aside from their insistence on individual optimizing foundations, there is often very little to distinguish them from non-neoclassical approaches, as will become obvious as we proceed. In many cases – theories of unemployment being the most important example – they were motivated by developments in alternative approaches and add very little to it except optimizing behaviour.

13 Many neoclassical economists, however, are willing to forsake perfect competition

in goods markets, but it can be shown that relaxing this assumption and assuming monopoly or monopolistic competition will not change the central conclusions of the model with variable coefficients.

14 See section 5.3 for a fuller discussion.

15 See Darity (1981) for a discussion of the difference involved and a synthesis.

16 See Burmeister (1980) and Marglin (1984a).

17 Given the so-called Inada conditions for existence.

18 The Inada conditions mentioned above will guarantee existence of long-run equilibrium in this model.

19 See Steedman (1977) and Steedman, Sweezy *et al.* (1981). More of this below. Believers in the labour theory of value often regard disbelievers as non-Marxists.

20 It should be noted here that in the one sector model the question of labour values and the question of the exogeneity of the rate of exploitation as opposed to the real wage are distinct issues, so that one can subscribe to the latter and drop the labour theory of value. In multi-sector models where the definition of the rate of exploitation could depend on the labour theory of value, this independence could break down.

21 Similar arguments about the social determination of a_1 can also be made. See Marglin (1984a).

22 See Elster (1983) for one view on whether Marx did, or should have, assumed fixed coefficients.

23 See Rowthorn (1980) for a fuller analysis of the Marxian theory of wages. In *Capital* Marx stressed the competitive aspects of wage determination, involving supply and demand forces, and did not analyse the role of trade unions. In his subsequent *Wages, Price and Profits*, the role of class struggle made an explicit entry with the analysis of trade unions, although their power was assumed to be regulated by supply-demand factors such as the size of the reserve army.

24 This model follows Findlay (1963) who actually considers a two-sector version which we shall examine in chapter 6.

25 Goodwin (1967) assumes that the output-labour ratio increases at an exogenously fixed rate, but, since we are abstracting from technical change in this chapter, we set this rate equal to zero.

26 See Goodwin (1967) and Gandolfo (1980) for an examination of the mathematics.

27 A large literature has evolved around this particular model. See, for example, Desai (1973), Shah and Desai (1981), Ploeg (1983), and Goodwin, Kruger, and Vercelli (1984).

28 Here cycles are obtained by endogenizing a parameter. Another possible way of obtaining cycles or instability is by examining the dynamics *behind* the equilibria, as in the neoclassical model of section 2.3.1.

29 See also Kenway (1980), Roche (1985), and Sardoni (1986).

30 Another method, used by Foley (1986), is to consider spending lags in the circuit of capital.

31 This is sometimes called Harrod's short-run problem to distinguish it from his long-run problem, mentioned above. Note that this problem is assumed away in the neoclassical model, which does not allow investment to be different from saving.

32 Some consistent theory of how ex-post saving and investment are equated in this short run must be supplied; an inventory adjustment could do the job.

33 See Baumol (1970), Gandolfo (1980), and Blatt (1983) for brief and simple descriptions of these developments using difference equations.

34 Several early models were developed by Kalecki (1933, 1935), which generated cycles as a result of lags in investment spending, and the fact that rising capital stock

(as a result of investment) reduced investment spending. While Kalecki's models did rely on the principle of effective demand and hence may be called neo-Keynesian, they clearly predated Keynes, and it would thus be inappropriate to call them 'neo-Keynesian'. In any case, we have named a separate type of closure after him. We will in fact argue that all the models of the type we are considering are more in the Kalecki-Steindl tradition than in the Keynesian tradition.

35 Chang and Smyth (1971) formalize the model to produce a limit cycle. Kaldor ignored class struggle effects emphasized by Goodwin (1967) (who, however, ignored effective demand). Skott (1989) introduces class struggle into a Kaldorian model by making the rate of change in output depend on conditions in the labour market.

36 Although, as we shall see later, the distinction between the two approaches is blurred with variable coefficients of production.

37 Forced saving plays a greater role in the *Treatise* (Keynes, 1930). Robinson (1962) discusses the difference between the two types of adjustment mechanisms for the short and long runs.

38 A variant in a discrete time model could assume $g = g(r_{-1})$, where the subscript -1 refers to the previous period. Such a model would allow actual and desired rates of accumulation to diverge in the short run, only to become equal in long-run equilibrium, in which $r = r_{-1}$.

39 We do not include the work of Kalecki, who is often bracketed with the neo-Keynesians, since one of our other closures is named after him.

40 Since Kaldor did not assume that the labour market cleared due to wage flexibility, but due to goods market adjustment through the investment function, his approach is not the standard neoclassical one.

41 See also Asimakopulos (1969, 1970).

42 Note that the neo-Keynesian model provides a different 'solution' to Harrod's long-run problem: by allowing for equilibria without full employment! The solution to Harrod's short-run problem is obtained by having the rate of profit as the independent variable in the desired accumulation function, and assuming that $g' < s$ (that is, a weaker investment response than the saving response, to changes in r).

43 See Dutt (1987b) and Amadeo (1989). Leijonhufvud (1974) later recanted.

44 Book length treatments by Harris (1978) and Marglin (1984a) devote a great deal of time to the three approaches. Harris incorporates Kalecki within the neo-Keynesian tradition, and devotes a passing footnote to Kalecki and Steindl in considering realization problems within the Marxian approach. Marglin believes that the Kaleckian approach is not possible within the one-sector approach using fixed coefficients. But, as argued in Dutt (1987a), the Kaleckian approach eludes him because he imposes – without acknowledgement – full capacity utilization.

45 Macro textbooks find hardly any reference to Kalecki's work, despite the fact that what goes as Keynesian theory is actually closer to Kalecki's work. See Sawyer (1985).

46 See Feiwell (1975). Bhaduri's (1986) recent macro text gives equal attention to Keynes and Kalecki, and uses an approach closer to Kalecki's. Sawyer (1982) argues that macroeconomics based on Kalecki's approach would be a superior type of macroeconomics. Sawyer (1985) offers a book-length treatment of Kalecki's work, and argues that the work of Keynes and Kalecki should be seen as distinct, despite their shared emphasis on aggregate demand. For a detailed analysis of Kalecki's microeconomic theory, and its relations to his macroeconomics, see Kriesler (1987).

47 Actually Kalecki distinguished between two types of sectors – fixprice sectors (with markup pricing) which primarily produced manufactured goods and flexprice (in which prices were demand determined) sectors producing mainly primary products and/or intermediate good. In our one-sector analysis we ignore the flexprice sector of Kalecki's analysis, which in any case played a much smaller role. Kalecki's analysis of the firms' behaviour went through several stages, and he was unable to solve several problems such as those of aggregation and how to define an industry. We do not go into the details of these developments, discussed by Kriesler (1987), because as Kriesler argues, Kalecki's macroeconomic theory (with given markup and demand-determined output) does not depend on how exactly the firm is modelled.

48 It has been argued that Kalecki's departure from competition also distances him from the Marxian approach; see chapter 7.

49 This dependence of investment on the stock of capital implies that expansion resulting from investment increased capital stock and eventually reduced investment; Kalecki's analysis of the business cycle often hinged on this assumption. He also examined the implications of lags in investment behaviour, which too, yielded cycles. We will abstract from these features of Kalecki's analysis, examining instead the steady growth models. As mentioned in section 3.2, these steady growth models can be extended to allow for cycles by including additional relationships in them.

50 Similar ideas can be gleaned from the later work of Baran (1957), Baran and Sweezy (1966) and Sylos-Labini (1956); see chapter 7.

51 Our theory of investment has been an aggregative one, given our interest in the representative firm. We have therefore not distinguished here between firm and economy-wide r and X/K. The question arises as to whether the firms respond to their own or the aggregate rates of profit and capacity utilization rates. If r captures internal finance effects à la Kalecki, the firm's r seems to be relevant, although the economy-wide one may also be, if it provides information on future profitability and hence future internal saving possibilities. If r captures expected profitability effects, the firm's r may be relevant if the firm bases its profitability estimates on only what is happening within the firm, and the general r may be relevant if it gives weight to what is happening outside it perhaps because it interprets its own performance to be excessively dependent on chance events. Regarding X/K, if the firm compares its own desired and actual capacity utilization rates in making its investment plans, the firm's own X/K is relevant; but if these plans depend on overall market conditions, the economy-wide X/K is the appropriate one to consider. Our analysis does not examine how total sales (production) are distributed between firms. Then if firms charge prices according to the markup formula, for firm j, $r_j = z_j/(1+z_j)(X/K)_j$, is true (where subscripts j refer to the variables and parameters for firm j). If, in the investment function, both r and X/K refer to variables for the firm, then since the firm may be assumed to be knowing that the above relation holds for it, it may be appropriate to assume that in $g = g(r, X/K)$, the second partial derivative is interpreted to be the effect of changes in X/K on g with r held constant. In that case, $g_1 > 0$, but g_2 cannot be signed definitely, since it reflects the effects of a rise in X/K and a fall in z (to keep r constant). And without $g_2 > 0$ the results of our model (especially that with a rise in z, g will fall) do not follow. This point has been raised by Marglin and Bhaduri (1986). If either one of r and X/K in $g( )$ is interpreted as macro variables, the relationship between r and X/K at the firm level need not be valid, and, though they are valid at the macro level, need not be taken into consideration by the firm.

52 See Bleaney (1976). The Kalecki-Steindl model formalizes Baran and Sweezy's

(1966) analysis quite faithfully; their potential surplus can easily be measured in terms of our model as the profits which would emerge if output were at full capacity, but with the same income distribution, that is, the surplus that would emerge if there was no demand problem. Baran and Sweezy acknowledge their debt to Kalecki and Steindl.

53 Amadeo's (1986) presentation explicitly distinguishes between the actual and desired (which he calls 'planned') degrees of capacity utilization, and his long-run equilibrium allows them to differ. Our own version makes the actual rate a variable to be determined, and our 'full' capacity can be interpreted as 'desired' capacity, so that here too, it is possible to argue, the two are different. But see below.

54 These points regarding cost curves were made by Sraffa in his 1925 paper, 'Sulla relazioni fra costo e quantita prodotta' (see Maneschi, 1986).

55 A plausible modification of this approach could assume that firms set the price as a markup on the 'normal labour cost' Wa_0^* which is assumed to be constant (see, Harris, 1978, Dornbusch and Fischer, 1978). This implies, at least in the short run, that even with variable coefficients, price would not change with output. However, in the long run, a given a_0^* would be difficult to justify.

56 With given z this implies that the real wage varies countercyclically, contrary to most empirical evidence. The normal a_0^* theory can rescue us from this problem. In the longer run, if z is variable, the real wage will not necessarily vary inversely with capacity utilization.

57 This is essentially a model of monopolistic competition, but without perfect entry driving profits to zero. If we assume a given number of firms, the model on the production side is similar to Weitzman's (1985).

58 Though Kalecki himself provided such interpretations in Kalecki (1939–40), he later gave up these attempts, no doubt realizing their futility. Kriesler (1987) correctly argues that these excursions were, for him, a digression.

59 For extensions along neoclassical lines using oligopoly theory see Cowling and Waterson (1976), Cowling (1982), and Sawyer (1985). Auerbach and Skott's (1988) critique of this approach does not apply to the second interpretation of Kalecki's approach (see next paragraph); it merely underscores the difficulties of the neoclassical interpretation.

60 See Bhaduri (1986). This can be justified using Simon's (1979) concept of bounded rationality.

61 Kalecki (1971a, 1971) explicitly described the role of trade unions in affecting the markup.

62 This is true with variable coefficients; the Marxian approach fixes a_1 by fixing the real wage and assuming all savings are invested, while the Kalecki-Steindl approach allows a_1 to vary and assumes the desired accumulation function.

63 See also Sherman (1987) for a discussion of Marx's analysis of realization problems, especially in the context of the business cycle. It should be pointed out also that the Kalecki-Steindl model does not necessarily imply long-run stagnation, which would appear to be too underconsumptionist for many Marxists, since the markup z could be changing cyclically for reasons we shall turn to later to generate cycles rather than a long-run crisis.

64 See Dutt (1984a) for a fuller analysis and references to the relevant literature. As noted above, the markup rate may have other determinants. The role of class struggle forces will be considered in chapter 4 and cyclical factors will be mentioned later in this chapter.

65 For a given z, the rate of growth of capital is equal to the growth rate of output,

given (2.15) and since r is constant in equilibrium. When z changes the two rates of growth will be different.

66 While the model can be extended in this way, we can capture these interactions (in the stable case) by studying the effects of variations of z in our basic model; for example, a downward shift of the $dz/dt = 0$ curve can be represented by a fall in z.

67 Effects relating to the power of unions will be considered in chapter 4.

68 For an example, see Taylor (1983), chapter 2.

4 Money and inflation

1 However, money and/or other assets are very much there in some of the models (albeit behind the scenes) and without them problems of effective demand would not arise. Note also that except with the artificial assumption of a fixed money wage our models do not solve for monetary variables.

2 We consider only 'helicopter' money, and do not link up money creation with government budget accounts. The analysis can be extended in this direction, following Taylor (1979, 1985).

3 An alternative assumption would be to make s depend on the ratio of real balances to capital, M/PK, a ratio which is directly related to the ratio of real balances to both real income and real wealth. See footnote 5 below.

4 See Stein (1971) for a derivation.

5 See footnote 3. In this case the model can be reduced to two equations given by

$$n = s(k)r$$

and

$$k = k(r + \hat{M} - n)$$

with two variables, r and k. The first equation implies an inverse relation between k and r and the second a positive relation. A higher $\hat{M}$ will shift the curve for the second equation to imply a higher k for a given r, so that k^* will rise and r^* will fall.

6 Real effects have been introduced into neoclassical models also by making real money balances enter the production function as a factor of production (Stein, 1971).

7 See de Brunhoff (1976) for an authoritative account and Roche (1985) for a simple exposition.

8 Marx actually determines prices in terms of labour values, but this can be thought of as a simplifying approximation under the assumption of equal organic compositions of capital.

9 If M is interpreted as the total amount of gold, changes in τ could be taken to reflect hoarding and dishoarding.

10 If there are decreasing returns in gold production, a higher $\hat{M}$ (representing a higher rate of gold extraction) could of course change the price. See Lavoie (1986).

11 This passiveness is a quantitative one – in the sense of the quantity of money affecting the price level. Qualitatively, money can be argued to play a much more active role in the Marxian system, by allowing income to be hoarded and not spent on goods, creating the problem of effective demand. See chapter 3.

12 See Foley (1983).

13 See Foley (1983). This approach is more popular among Kalecki and the post-Keynesians; see sections 4.3.1 and 4.4.1 below.

14 We now need to distinguish g^d (the *desired* rate) from g (the *actual* rate) since the two

could differ. It may have been preferable to assume that desired accumulation depends on $\hat{P}$ (capital gains), an assumption that will be incorporated in some later models, but is left out here for simplicity.

15 As before, Ω (with subscripts where necessary) will be the adjustment coefficient for each of the models.

16 We do not consider the popular textbook model which explains inflation in the short run by a Phillips curve, since in the long run it is completely neoclassical and thus no different from the model of section 4.1: the economy grows at the rate of growth of labour supply, and the (exogenous) rate of growth of money supply determines the rate of inflation.

17 While we do not explicitly introduce an interest rate into the models, we could assume it to be fixed by the monetary authorities or by the actions of speculators, depending on what we think is the mechanism which makes money supply endogenous.

18 These models, however, differ from the following one because they allow for variable coefficients of production, exogenous money supply growth, and a more detailed treatment of asset markets.

19 The level of the desired real wage, which represents the target that workers wish to achieve, depends on the objective conditions of the labour market and the economic environment in general, in which the workers find themselves. This implies that workers will not desire the moon.

20 If we make $\tilde{V}$ a function of the unemployment rate, to incorporate the idea that higher unemployment rates weaken the bargaining power of workers, the equation would be similar to the Phillips curve. The inflation term could reflect both indexation or wage push due to inflationary expectations where expectations are, for simplicity, rational.

21 If $\hat{P}$ did not enter the g^d function the IS curve would be vertical.

22 We do not examine the dynamics of this model in more detail. Alternative assumptions regarding dynamics could produce different results.

23 The statement would be exactly true if $\hat{P}$ did not enter the desired accumulation function.

24 This model follows Marglin (1984a, 1984b).

25 The $\hat{P}$ term is omitted for simplicity; its inclusion would not change the conclusions significantly.

26 See the discussion of endogenous money in section 4.3.1. For a summary of Kalecki's views, see Sawyer (1985).

27 Kalecki also argued that, if the demand for money increases more than the supply of money and the interest rate increased, this will not have much impact on the rate of investment or savings, so that avenues of adjustment through their variation are blocked.

28 Note that (4.12) has already been incorporated into it, because we do not distinguish between g and g^d.

29 If we ignore the capital gains term from the investment function we get a vertical IS curve, so that g^* is determined only by 'real' forces, while $\hat{P}^*$ is determined by the rate of change in the money wage.

30 Such changes can also shift Ω_1 and Ω_2, the effects of which can be analysed in virtually the same way.

31 This result holds assuming that there are no further consequences for any of the other parameters, for example, those of the investment function, due to greater uncertainty, or 'capital strike'.

32 This is in the spirit of Post Keynesian writing on wage flexibility and unemployment. See Dutt (1986–7) and the literature cited there.

5 Technological change

1 Solow (1957) fitted some specific forms of neoclassical production functions to US data for 1909–49, and found that 87.5 per cent of the increase in gross output per man was attributable to technological change and the remaining 12.5 per cent to capital accumulation.

2 See Hacche (1979) for a simple exposition. The stability of this growth equilibrium requires the production function to have an elasticity of substitution of less than one. See Drandakis and Phelps (1966).

3 For models with expenditures on research and development, and education, respectively, see Phelps (1961) and Uzawa (1965).

4 For open economies an additional channel is the transfer of foreign technology.

5 This is apart from such issues relating to departures from optimality, and from the specific assumptions which are made to incorporate research and education.

6 This characterization follows Levhari (1966), Sheshinski (1967), and Hacche (1979). An alternative formulation, due to Bardhan (1970), measures learning with cumulative output rather than cumulative investment. The specific log-linear form is used for simplicity; any other form resulting in an upward-rising relationship between g and h would have sufficed. Arrow (1962) found microeconomic empirical support for the log-linear form.

7 Elster (1983) argues that Marx was wrong to assume this. He does agree, however, that for a given output level this is what Marx in fact assumed, although with changes in output the fixed coefficients technology became more capital intensive. Elster seems to believe that this was with given technology and with changes only in scale, but it is not clear that Marx was not talking of technological change, or in fact would want to distinguish between changes in technique (given technology) and technology.

8 In standard Marxian presentations, the wage fund has been included as capital, and the rate of profit has usually been expressed as a ratio of a flow to a flow. In circulating capital models it does not matter whether the denominator is a flow rather than a stock, but in our model it does. We follow our usual convention of expressing the rate of profit as the return on the stock of capital which does not include the wage fund. Harris (1981) presents a dissenting view, while Laibman (1982) provides some support for the assumption. Note also that equation (5.9) expresses the rate of profit as a ratio of quantities in terms of the good, and (5.9″) in terms of labour values. For our one-sector model, there is no difference between r computed in these alternative ways, but this is not generally the case.

9 See von Bortkiewicz (1906–7), Robinson (1942), Sweezy (1942), and Rosdolsky (1977).

10 It should be noted that, although Marx (*Capital*, Vol. III) believed that the law of the tendency of the rate of profit is 'in every respect the most important law of modern political economy', he did make it clear that it was 'only a tendency, like all other economic laws'. He devoted a whole chapter of *Capital*, Vol. III (Chapter XIV) to discuss 'countervailing influences' which worked in the opposite direction. Some of these are also discussed in *Grundrisse*. See also Rosdolsky (1977).

11 See, Okishio (1961, 1977), Roemer (1977, 1978, 1979), Van Parijs (1980).

12 See Roemer (1979, 1981) for a demonstration in a multisector economy. Laibman (1982) uses a similar graphical technique for a two-sector model.

13 For a model with different assumptions, see Glombowski (1983). The rate of exploitation is assumed to depend negatively on the rate of growth of employment. Assuming an increasing organic composition of capital (defined as the ratio of dead labour to living labour) at an exponential rate, Glombowski shows that the Marxian results of the falling rate of profit, rising unemployment, and rising rate of surplus value, are obtained. The analysis also shows that, if the assumption regarding the rising organic composition of capital is replaced by that of neutral technical progress, all this changes, underscoring the importance of the former assumption in generating the Marxian results.

14 See Hacche (1979) for a fuller discussion.

15 Harcourt (1963) argues that the validity of Kaldor's argument also depends on his notion of the representative firm. He shows, using a two-sector short-run model, that equilibrium with the full employment of labour requires specific and different types of pricing policies to be adopted by firms in the two industries, so that, if the economy has one representative firm, full employment cannot be ensured. While Harcourt's argument may be valid, Kaldor's results do not require full employment in short-run equilibrium.

16 It will not also do to assume (as Auerbach and Skott, 1988, suggest) a higher investment responsiveness in the long run and a smaller one in the short run, so that we get Keynesian output adjustment in the short run and full capacity in the long run. For if this were the case, a short-run contraction would push the economy to a downward Harrodian knife-edge, rather than to full capacity.

17 Kaldor assumes a linear desired capital-output ratio function which gives him a non-linear investment function which results in multiple equilibria. We abstract from such problems simply by assuming a linear investment function.

18 In section 2.3.3 we saw that given a_0,

$$g = [s\alpha + (s/a_0)(N/K)]/(s - \beta)$$

which must equal n, so that

$$N/Ka_0 = [n(s - \beta) - s\alpha]/s,$$

a constant, implying that

$$g = n + \hat{a}_0,$$

which from (5.16) implies $\hat{X} = g$, or the constancy of X/K.

19 In equilibrium, the rate of profit does not change, the real wage rises over time at a rate equal to the rate of technological change, and the distribution of income is constant. When technological change is not Harrod-neutral, since a_1 affects g, these conclusions cannot be drawn.

20 Eltis's (1973) model is different from this one. In his non-vintage model he has a desired accumulation function and a saving function similar to ours, except that the former has the rate of profit and the interest rate as arguments. While capacity utilization varies to bring saving and investment to equality at full employment as in our model, the interest rate does the job in his. However, the two models produce the same general result.

21 If we are in the capital-constrained case, in terms of Figure 4.4, the AB line will be vertical, with $gK/X = s(\alpha a_1 + \tau)/(s - \beta)$. An increase in α will shift this curve to the right, increasing $-\hat{a}_0$; this will increase g, but the higher $-\hat{a}_0$ has no role to play in this, since g is given by (2.34).

22 There is one negative influence of technological progress which has been hidden in the investment functions (2.14) or (2.31). If the rate of investment depends on an index of capacity utilization, in the presence of technological change, it is preferable to write it as

$$I/K = \alpha + \beta r + \tau' a_1 (X/K)$$

where $\tau = \tau' a_1$ has been substituted into (2.1), because the ratio of actual to potential output (which ought to be a measure of capacity utilization) is $a_1 X/K$, not X/K. In this case, if technological change implies that a_1 falls through time, the utilization of capacity will fall and hence the growth rate of the economy will decline. This strange result can be taken to indicate that technological change is usually Harrod neutral. This possible negative effect of technical change on investment will henceforth be ignored.

23 Since the real wage, however, increases, the distribution of income will be unchanged (as long as z is fixed).

24 See also Kalecki (1941). Kalecki's ideas are also spelt out in one of his last papers, Kalecki (1968).

25 The effects of these innovations can also operate through upward shifts in consumption functions. In the neo-Keynesian and Kalecki-Steindl models the result (through a fall in s) will be the same as that of upward shifts in investment functions. Baran and Sweezy are actually arguing that technological change will not push demand up enough to eliminate excess capacity; they should therefore not be interpreted as denying the weaker claim that they do not raise aggregate demand, even for non-epochal innovations.

26 The previous paragraph implies that this can happen even when investment responds positively to the rate of technological change.

27 Marx (1894) wrote that '[t]he falling tendency of the rate of profit is accompanied by a rising tendency of the rate of surplus value, that is, the rate of exploitation'. We have seen in chapter 3 that the rate of exploitation, ϵ, is the same as the mark-up, z, in our one-sector framework.

6 Two-sector models

1 For a discussion of other systematic differences between sectors, and a survey of models which take these sectoral differences into account, see Dutt (1989d).

2 In the literature the term long-period position is usually used for this type of equilibrium, but we use the two terms synonymously.

3 The models are thus in the tradition of those examined by Von Neumann (1945–6), Brody (1970), Abraham-Frois and Berrebi (1979), and others.

4 This is (2.6) of chapter 2; we will rewrite equations where necessary, renumbering them, to allow the mathematical structure of each chapter to be reasonably self-contained.

5 As shown above, these curves do not necessarily have to be concave to the origin.

6 An alternative formulation of the model can assume a fixed rate of exploitation, or share of profits in the total value of production, but is not pursued here.

7 Sardoni's analysis is valid, but only considers changes from one period to another; it is hence not a dynamic equilibrium in the sense of being self-repeating.

8 For other two-sector models introducing markup pricing, see Harcourt (1965), Mainwaring (1977), and Taylor (1983, 1984). Harcourt considers a very general model (with technical substitution, and endogenous markups depending on demand levels – although not in the sense of the neo-Keynesian models), but

confines attention to the determination of employment and distribution in the short run only. Further, his model differs from ours in including neo-Keynesian investment functions. Mainwaring also confines attention to the short run. Taylor considers long-run dynamics and equilibrium in addition to the short run, but uses neo-Keynesian investment functions. See also Bhaduri and Robinson (1980).

 9 To analyse them it is necessary to examine the dynamic process leading to the long-run equilibrium, allowing for differential rates of growth, and also possibly differential rates of profit. A simple convergence process will be formally considered later in section 6.2.3, but for now we restrict ourselves to long-run equilibria. Our analysis there may, but will not generally, imply equalization of rates of profit in long-run equilibrium.

10 There has not been much discussion of asset markets in models of this type. But see Panico (1980, 1985).

11 See section 3.4 for the discussion of other possible mechanisms and why they do not necessarily ensure full capacity. Kurz (1986) tries to provide a mechanism by which competition leads to 'normal' capacity utilization in long-run equilibrium in the classical sense. He argues that firms can choose between different degrees of capacity utilization, which will (due to overtime pay, for example) imply different unit costs. Profit-maximizing firms will thus choose the lowest cost utilization rate for given data just as they choose among alternative techniques. The forces of competition which drive profit rates to equality will drive firms to normal capacity utilization with lowest costs. The problem with this, as Kurz (1986) is aware, is that it is valid only in a situation of what he calls 'free competition'; while this is not defined precisely, the notion requires atomistic firms, which is not implied by classical competition (see chapter 7). While in Kurz's framework firms choose their rate of capacity utilization and they can sell all that they produce, in our oligopolistic environment, excess capacity exists due to the lack of aggregate demand at prices which the sellers fix.

12 The model perhaps makes more sense if we assume that profit rates are not equalized. Here, as in the rest of this section, profit rates are assumed to be intersectorally equalized, to set the stage for the discussion with differential rates of profit in section 6.2.4.

13 These equalizations can be thought of as being the result of the process described in section 6.2.2.

14 If the capital intensities are the same, the *g* and *n* curves will be parallel, and we have Harrod's long-run problem.

15 Mukherji (1982) produces the same result. Note that the dynamics described here does not necessarily imply full employment at long-run equilibrium, only that the unemployment rate becomes constant.

16 See also Mukherji (1982).

17 See Flaschel and Semmler (1986a).

18 See also Franke (1986), Hosoda (1985), and Kuroki (1986) for similar results in slightly different models.

19 More than one consumption good must thus be considered in the model.

20 The qualification 'almost' is introduced because Nikaido only has circulating capital, whereas the neoclassical models have fixed capital, so that exact comparison between the models is not possible.

21 We have found a similar condition for stability in a neo-Ricardian model of section 6.2.1, which could embarrass non-neoclassicals; it is required for convergence in a model in which the wage adjusts, rather neoclassically. The Nikaido condition is

required for the fixed wage model for adjustment to the prices of production, and could thus appear even more embarrassing to non-neoclassicals.

22 This follows Dutt (1988d).

23 Most models assume that output adjustments depend on profit rate differentials. In our model this assumption is equivalent to ours, since capital-output ratios are fixed.

24 Several authors, including Nikaido (1983, 1985) and Kuroki (1986) have considered non-market clearing rationing models, thereby formalizing the 'cross dual dynamics' process (see Semmler, 1986, Introduction). We assume, for simplicity, that the price adjustment is much faster than capital adjustments, so that we can think of the former being completed in the short run, and that the process involves the exchange of information in the nature of a tatonnement. The results obtained by Nikaido and Kuroki, in any case, show that the market clearing and non-market clearing models have identical stability conditions. Further, non-market clearing rationing models are difficult to interpret for certain types of competitive models, that is, those assuming price taking behaviour, as discussions related to the fixed price disequilibrium microfoundations models have brought out. An alternative to rationing is to allow for inventory adjustments. While this too presents difficulties for interpretation in models of price-taking behaviour, the stability properties change, as mentioned above.

25 Subject to the short-run existence condition being satisfied. This condition can be restated to say that the rates of profit are positive in each sector.

26 There is at least one contribution, Harris (1981), which suggests that wages should be considered a part of capital.

27 See Laibman (1982).

28 There already exist some attempts to introduce non-competitive price setting behaviour explicitly. See, Boggio (1986) and Dumenil and Levy (1986), which make rather complicated assumptions about pricing and production behaviour. The analysis here follows Dutt (1988d).

29 This is different from the D of section 6.2.2.

30 For example, we may have

$$g^2 = f(X_1/K_1, X_2/K_2)$$
$$g^1 = g^2 + \pi(r_2 - r_1)$$

with $\pi < 0$. It is assumed here that investment in the capital goods sector responds to generalized excess capacity in the economy, and the difference in the two sectoral growth rates responds to the profit rate differential. The second equation forces rate of profit equalization at steady state growth. Other models which have investment functions with profit rates and outputs as arguments make some similar assumptions to enforce rate of profit equalization in long-run equilibrium. Kuroki (1986) makes peculiar assumptions by imposing that the expectations governing investment decisions must respect the economy's budget constraint, which is not true for demand-determined models. Dumenil and Levy (1986) obtain profit rate equalization with what are in effect identical investment functions for different sectors, and requiring 'normal' capacity utilization at equilibrium, although they argue (later in their paper) that an equilibrium with excess capacity is also possible in their model.

31 See Fujimoto and Krause (1986) for an analysis of technical progress in a model showing convergence to the prices of production with equalized rates of profit.

32 Changes in the markups over time could be examined by assuming that they fall

when excess capacity exists. While there may be reasons why firms reduce their markups when they observe excess capacities, it is by no means obvious that the aggregate markups in the two sectors must fall with excess capacity, thereby taking the economy to full capacity and perhaps (with some restrictions on our investment functions) a situation of equalized profit rates. First, with excess capacity and unemployment, the bargaining position of workers may be low, resulting in a reduction in the real wage, implying a rise in the markup in the consumption good sector. Secondly, with low rates of growth and capacity utilization, the degree of concentration could increase in the industries, creating pressures for an increase in markups. Thirdly, high fixed costs (with excess capacity) could increase markups along the lines argued by Kalecki (1971).

33 This is adapted from Taylor (1979).

7 Some doctrinal issues concerning two-sector models

1 See Pasinetti (1977) and Abraham-Frois and Berrebi (1979) and Mainwaring (1984) for expositions of the Sraffian system. Sraffa's work is similar to the work of Dmitriev (1974) and Schwartz (1961).

2 He also develops the concept of the standard commodity which, if taken as the unit of measurement of wages, implies a linear relation between the wage and the rate of profit. While the standard commodity may be taken to make the relationship between wage and profit more transparent by making the relation linear, we shall not be concerned with it since it contributes nothing to our analysis of growth and distribution.

3 Hahn does admit that Sraffa's analysis does imply that the question of the *stability* of the neoclassical model is not as simple as one might have thought from the one-sector homogeneous capital model. However, this is not an internal inconsistency of the model, but its failure to show that the system is stable (unless this is taken to be a criterion of consistency). Also, Sraffa type systems, as already discussed above, have their own stability problems.

4 See Robinson (1953). For an authoritative account of the debates in capital theory, see Harcourt (1972). For a brief treatment of neoclassical capital theory see Dixit (1977); see Bliss (1975) for a more extended discussion.

5 See Hausman (1981) and Dow (1985) for the background to the following comments.

6 See Sen (1978).

7 See Garegnani (1976), Milgate (1979, 1982), and Kurz (1985).

8 Doing so implies giving a balanced-growth interpretation to Sraffa's work which is not necessarily implied in Sraffa's writings. The interpretation follows Abraham-Frois and Berrebi (1979), Schefold (1980), and Mainwaring (1984), among others. Sardoni (1981) has criticized this interpretation of Sraffa's analysis. Most of our comments in this chapter, however, do not depend on the balanced growth assumption.

9 When Sraffa does consider durable capital goods, he treats them as joint products, so that the relevant input-output ratios are written on the left-hand side of the equations. We write them on the right-hand side, without changing anything of substance in this regard.

10 See Roncaglia (1978), Bharadwaj (1978), Eatwell (1977), and Garegnani (1983, 1984). This interpretation of the classical economists is not unanimous. Marshall

and Jacob Hollander viewed the neoclassical general equilibrium system as one which developed from the Ricardian tradition. Samuel Hollander (1979, 1985) is the most outspoken supporter of this viewpoint; he sees Ricardo and the neoclassicals both as trying to explain the determination of prices, distribution, and outputs together. For a critique of Hollander's view, see Roncaglia (1982).

11 See also Coddington (1983).

12 See Keynes (1936, 297) for another good example.

13 We therefore argue that the distinction that Schlicht (1985) draws between substantive and hypothetical isolation is too blurred to be of any significance in actual use.

14 See also Coddington (1983, 21).

15 See Coddington (1983) for further discussion of the last two points. The interpretation of Keynesian economics that finds the dichotomy with wages and prices fixed is a popular, but incorrect interpretation of Keynes's own work; it is, however, a correct one for much of what passes as Keynesian economics.

16 This may not be a valid characterization of all of 'classical' theory but is Keynes's view on those classical economists he was taking to task. Another example is Marx's dichotomy (recently examined by Negishi, 1985) between the theory of distribution between capitalists and workers, and the theory of relative prices, which determines distribution between capitalists.

17 Our discussion does not depend on whether it is actually possible to write down the entire set of equations. In fact, if it were possible, there would not be much need for using analytical dichotomies!

18 Following Negishi (1985), suppose there are two vectors x and y, which can in general be determined by the two sets of functions

$$F(x, y) = 0 \qquad\qquad\qquad\qquad\qquad\text{(a)}$$

and

$$G(x, y) = 0 \qquad\qquad\qquad\qquad\qquad\text{(b)}$$

If we can show that condition (a) is actually

$$F(x) = 0 \qquad\qquad\qquad\qquad\qquad\text{(c)}$$

then there is a valid dichotomy in the sense that the determination of x can be studied quite independently of what happens to y. Keynes's 'classical' economists were assuming a dichotomy in this sense, where x is the vector of real variables, and y money and other nominal variable. An interpretation of Keynes would be to argue that (c) is not true: thus changes in the components of y would have effects on x.

19 No other rationale for this dichotomy is adequate, since what is considered at a point in time to be non-economic is purely arbitrary, determined by the chance developments in the progress of the subject up to that time.

20 See Schlicht (1985) for a more complete analysis of this type of a dichotomy, which he calls temporal isolation. In this book we use the terms 'period' and 'run' synonymously, though in some contexts a theoretical distinction between the two terms can be drawn.

21 See also Garegnani (1983).

22 See, for example, Levine (1974, 1975, 1977, 1985), Burmeister (1975, 1977), Eatwell (1977), Roncaglia (1978), and Mainwaring (1979) for the two sides to the debate on this issue.

23 This problem was raised in Harcourt and Massaro (1964). For a formal analysis, see Mainwaring (1979).
24 See Roncaglia (1978) and Bharadwaj (1978). See also Harcourt (1981) for a discussion of alternative notions of centres of gravity.
25 Sraffa's framework is actually supposed to strengthen Keynes's argument that saving and investment cannot be brought to equality by variations in the rate of interest, as in the models Keynes called 'classical'. The 'classical' theory depends on the existence of an inverse relation between investment and the rate of interest, a function which cannot necessarily be derived in a world in which 'capital' is not one factor of production. Since the interest rate cannot be relied upon as an equilibrating variable, output variations have to bring saving and investment to equality.
26 If the input-output ratios and one of the distributional variables, were observed, the prices of production could be solved for. This is what we earlier referred to as theorizing without counterfactuals. Such a procedure obviously does not imply that the equations 'determine' prices. See Sen (1978).
27 Mainwaring (1979) has shown that, if such interrelations are taken into account, the wage-profit frontier can have a positive slope.
28 'Competition' has other roles than those discussed here, such as enforcing minimum-cost production and generating technological change. See Harris (1988) for a broader analysis, and Dutt (1988e) for a more detailed discussion of the issues raised here.
29 Several writers have remarked on the neglect of the role of the firm in classical theory. See Levine (1980), for example. This should not be taken to imply that Marx did not have a theory of the firm in other contexts such as the organization of production; only that the firm is hidden from view in the analysis of capital mobility.
30 See Arrow (1959) and Fisher (1983). This is not to imply that it is not possible to rigorously formulate assumptions which would generate a process resulting in an equilibrium approximated by perfectly competitive equilibrium; a model of monopolistic competition, for example, could supply these assumptions.
31 Roemer (1981) has shown that in a generalized Marxian economy with two classes (workers and capitalists), an equal rates of profit equilibrium with neoclassical assumptions of competition (including price-taking behaviour and profit maximization) can be shown to exist if capitalist firms have access to a perfect credit market. This shows that the equalization of rates of profit can be given a completely neoclassical interpretation. However, it does not show that price-taking behaviour is necessary for classical competitive equilibrium.
32 See Foster and Szlajfer (1984) for a collection of writings by theorists in this tradition.
33 This has subsequently been formalized. Semmler (1984, 157–8), for example, takes exogenously given, different, profit rates for each industry instead of an uniform rate. Kotz (1982, Appendix 2) uses the same approach, but explains the higher profit rate in monopolized industries (compared to the competitive uniform rate) in terms of the height of entry barriers.
34 Shaikh (1978, 1980, 1982) has criticized also the neo-Ricardian school for confusing the classical and the neoclassical notions of competition. While such a criticism is valid for some of those in the neo-Ricardian tradition who have been discussing the choice of technology issue and its relation to the tendency for the rate of profit to fall, the concept of competition employed in Sraffa is essentially the same as that employed in Marx.

35 These issues came up earlier in the exchange between Whitman (1942) and Kalecki (1942).
36 See Cornwall (1977), Pasinetti (1981), and Harris (1985).

8 Alternative models of North-South trade

1 There have, of course, been many contributions in the formal trade theory literature which have assumed various departures from an undistorted perfectly competitive economy, and indeed, these contributions have shown the possibility of losses from trade. The approach, nevertheless, has been that of finding optimal policies to get rid of the distortions and imperfections, and reaping the benefits of trade – implicitly assuming that the optimal policies are what governments will want to be able to pursue.
2 This is not to imply that models with completely neoclassical assumptions cannot be constructed to study North-South interaction. See, for example, Jones (1965, 1971), a variant of which is used by Darity (1982b) in his analysis of the eighteenth century Atlantic slave trade.
3 The literature on North-South models is substantial and multiplying rapidly, and we shall here be concerned with only a small subset of it. The structure of many of the North-South models that have been developed is different from the one considered here. Those of Bacha (1978), Chichilnisky (1981, 1984), Spraos (1983), and Dixit (1984) are not dynamic in the sense of studying accumulation patterns over time; and Darity's (1982a) does not explicitly introduce details regarding the terms of trade and the saving-investment process. Krugman (1981a) and Dutt (1986a) deal with technological change and do not assume a given pattern of specialization as will be assumed in this chapter. (See chapter 10.) Brewer (1985) differs from the models we look at here in not allowing for fixed capital (capital is only a wages fund), and in allowing the international mobility of capital. Capital mobility is also introduced by Burgstaller and Saavedra-Rivano (1984). We will not introduce capital mobility until section 8.4. For a survey of these and other North-South models, see Dutt (1988c).
4 The analysis follows Dutt (1989a).
5 Some of the subsequent chapters also examine the dynamics for particular models.
6 Given our steady state reference, in equilibrium, the North and the South have to grow at the same rate. If both regions are neoclassical, and they have given rates of growth of labour supply, a steady state is impossible unless the rates of growth of labour supply accidentally coincide. If both countries are neo-Keynesian, their rates of growth will be determined solely by their saving patterns and the nature of their investment functions, and these, except accidentally, will not be equal. Similarly, a neoclassical economy and a neo-Keynesian economy cannot be combined in the same model. It should be made clear that some of these models are ruled out because of the assumptions made in our general framework; they can become possible if modifications in it are made. For example, Conway and Darity (1985) allow for a long-run equilibrium where both the North and the South are neoclassical and have different rates of growth of labour supply. This is possible because, unlike what is postulated in our general model, they assume non-constant returns to scale (the North exhibits increasing and the South decreasing returns to scale), and allow factor substitution in production.
7 This is argued by one of the main contributors to the informal uneven development literature, Amin (1976).
8 Michaely (1984) calculates the income levels of exports and imports (by weighing per-capita incomes of countries by their shares in world exports and imports of

a commodity) of commodities and finds the dispersion of these indices to be small *between* the (nine) major SITC categories and high *within* each category. Even if we find a very fine commodity classification, it is likely that physically similar commodities, in terms of how they are perceived by consumers in the North and the South as regards their degree of substitutability, are different depending on whether they are produced in the North or the South. All this suggests that it is not clear that the assumption that there are two goods produced by the two regions is better than the assumption that there are two composite goods produced one each by the two regions.

9 Although we use the same symbol k as in chapter 6, it refers now to something different; there is no reason, however for confusion.

10 (8.11) can be derived from these five.

11 We could examine the dynamics out of long-run equilibrium using the dynamic equations $dK_i/dt = g^i K_i$, replacing (8.18) by the condition that K_i are given in the short run, but otherwise maintaining the other equations for all runs. Other models can have different equations for the short run and the long run, as in Conway and Darity (1985). There is no guarantee, of course, that the long-run equilibrium for any particular model would be stable.

12 Existence conditions can be derived as in earlier chapters, and are all assumed to be satisfied in what follows.

13 Entries with '?' refer to cases in which the sign is ambiguous.

14 It can be shown that with an additional desired accumulation function the model will be overdetermined.

15 This is the main, although not the only, definition of uneven development that we will adopt in this book. For a different definition, in terms of real wages, see chapter 10.

16 This, and Baran's work, is discussed in more detail in Dutt (1988a).

17 Or allow for capital mobility, as does Emmanuel. See also our comment on Braun, below.

18 See, for example Emmanuel (1972), Amin (1976), Gibson (1980), Evans (1976), Braun (1984), and Brewer (1980, 1985). Emmanuel's initial presentation was in terms of labour values.

19 See Singer (1950) and Prebisch (1950). See Haberler (1961), Findlay (1981), and Spraos (1983) for excellent surveys, and Sarkar (1986) for a recent statistical evaluation.

20 For the short run, we could take fixed values of k and f, due to the fact that K_n, K_s and K_f are fixed in the short run. These would replace the following two equations, and we could proceed otherwise in the manner described in what now follows. The following two equations must be satisfied if we are to solve for k and f.

9 Endogenous preferences and uneven development

1 Adverse movements in the terms of trade have, of course, been supposed to have been caused by a large variety of factors, of which low income elasticity of demand is just one. See, Haberler (1961), Findlay (1981), and Spraos (1983) for excellent surveys.

2 That the South imports all investment goods is not crucial for our analysis. What is crucial is that at least a core is inelastically required. While we assume away the importing of Southern intermediate goods by the North, there should be no difficulty in modelling that phenomenon along the lines pursued here; taste shifts would be replaced by technology shifts (reflecting the emergence of synthetic substitutes, for example).

3 The role of W_n is very much like that of a numeraire; the fixity assumption is therefore not a crucial one.

4 The Kalecki-Steindl model makes investment depend on the rate of profit as well as the rate of capacity utilization, while our present version makes it depend only on the rate of profit. Equation (9.2) can be interpreted as a reduced form of the general investment function after substitution from (9.3); z is incorporated into the form of the investment function. In this chapter we will not be changing z, so that we need not fear forgetting the effects of changes in z on g_n.

5 We therefore have a two gap model in which the two gaps become equal given the assumption that all investment goods are imported.

6 The existence of equilibrium requires that the saving and investment schedules intersect, and that they do so at a level of capacity utilization which does not require more capital than is available. The values of the parameters of the model will have to satisfy some conditions, similar to the ones discussed in earlier chapters. The stability of this model in the short run would require, whatever the mechanism assumed for I_s, that the responsiveness of savers to changes in output in the North is greater than the responsiveness of investors to the same.

7 A change in W_n will affect the equilibrium values of none of the real variables in the model; it will simply change, equi-proportionately, P_n and P_s. Thus there is no need to assume its level fixed except to pin down the values of the nominal variables of the model.

8 It appears that this definition of uneven development is different from, and in fact contradictory to, the definition given in the previous chapter. By the present definition, long-run movements (and the stability of long-run equilibrium) are taken to be signs of the evenness of development, while in the previous definition changes in the long-run equilibrium level of k due to parametric shifts determined evenness. It will be clear from our subsequent discussion, however, that the two definitions are equivalent.

9 The analysis follows Dutt (1988b).

10 This is not Prebisch's only reason for terms of trade deterioration. See also Flanders (1964) and Sodersten (1971) for discussions of this aspect of Prebisch's analysis.

11 But not impossible. The intercept terms of consumption functions (of commodities) can be tied to levels of capital stock, as done in Pasinetti (1981) and Taylor (1984); this assumption will make consumption behaviour mimic Engel's law and also yield steady states.

12 See, for example, Kindleberger (1956), who makes this distinction, and points out that, while there is no support for the proposition that the terms of trade between primary and industrial products has moved against the former, there is some support for the proposition that the terms of trade for less developed countries have done worse than those of more developed countries.

13 See Haberler (1961).

14 Singer compares his current views to his previous views as follows: 'Singer I assumed the central/peripheral relationships to reside in the characteristics of different types of *commodities*, i.e., modern manufactures versus primary commodities. Singer II now feels that the essence of the relationship lies in the different types of *countries*.'

15 One could formalize this argument using Krugman (1981b) type models with differential products, for example.

16 There could be certain goods where the opposite could be true. But the class of such goods will probably include only exotic handicrafts and knickknacks. For our

purposes it makes no difference whether there are actual quality differences or perceptions regarding the same.

17 It may be objected that this kind of formalization is unsatisfactory because taste changes are simply assumed, not derived from more 'basic' changes of quality, varieties of goods, and expenditures on advertising. It should be understood that these explanations are merely meant to be interpreted as illustrative stories, and our analysis will not be affected by what exactly that story is. Secondly, we are interested in the macroeconomic *effects* of such changes, and there seems to be no reason that such effects would depend on how exactly these microeconomic stories are formalized. (9.16) is therefore assumed as a 'primitive'.

18 If the North is growing (as it will in our model), there will be no way of empirically distinguishing between non-homothetic tastes and taste changes in our sense from income-consumption data. However, there are important theoretical differences between the two types of formalization, which, apart from the reasons already given up, would make ours more preferable. First, in our formalization, shifts in expenditure shares occur only in the long run, not in the short. Secondly, shifts in consumption patterns can occur even without changes in income levels per capita in our formalization.

Our formulation can also be modified to allow for Duesenberry-type ratchet effects. If taste changes of the type which increase a are irreversible, then for any k, a rise in k will increase a, but a fall in it will have no effect. This will make it more difficult to reverse the forces of uneven development, as discussed below.

19 Early discussions are found in Nurkse (1953) and Myrdal (1956). A good survey is available in Kottis (1971).

20 In India, for example, despite prohibitions on imports of video recorders, such products have been imported as 'parts' and sold after 'processing' which in many cases merely involves the fixing of a new product sticker!

21 Notice that these international demonstration effects are only supposed to have a role on Southern capitalists. This is a reasonable assumption, given that the Southern rich are exposed to such effects, and they are in a position, economically, to be affected by the imitation effect. In reality, there are some spread effects to lower income groups. (For example, the preference for audio tape recorders shown by low income office guards in Calcutta.) But the real imitation is in the higher income brackets. (Notice the proliferation of video-tape recorders among the rich in India, following that among Northern consumers; many Southern consumers would consider the ownership of that gadget as a 'status symbol', providing some support to the imitation effect, as contrasted with the pure information aspect of the international demonstration effect.) Possible lags in the generation of these effects are ignored in our formalization, for simplicity. We can modify our analysis by allowing ratchet effects; increases in a will increase b, but nothing can reduce it. This will happen if the taste changes for Southern capitalists are irreversible.

22 This will be the shape, for instance, if (9.16) and (9.17) take the forms

$$a = Ak \quad \text{for} \quad k < 1/A \quad \text{and} \quad = 1 \text{ if } k > 1/A$$
$$b = Bk \quad \text{for} \quad k < 1/B \quad \text{and} \quad = 1 \text{ if } k > 1/B$$

and s_s is constant. In this case,

$$e_a = Ak/(1 - Ak) \quad e_b = (1 - s_s)Bk/[s_s + (1 - s_s)Bk].$$

Low values of k make both positive elasticities small, and hence will imply, by (9.19) that $dg_s/dk > 0$. As k increases, e_a and e_b will rise, and eventually make $dg_s/dk < 0$

when their sum exceeds 1. When $k > 1/A$, $(1-a) = 0$, so that $g_s = 0$. The curve is not drawn for that state as A is assumed to be very small.

23 Since in reality both terms of trade deterioration and uneven development can be caused by a variety of factors not considered in our model (see the previous chapter) this result obviously has no implications for actual historical tendencies.

24 Such a policy change may also have detrimental efficiency effects which could be captured in multi-sector models, but which could be portrayed here as changing the technological coefficients. We abstract from such effects in the discussion which follows.

25 If one incorporates the irreversibilities of the types mentioned in earlier footnotes, it will be harder to reverse the process of uneven development.

26 We are comparing the effects of policy changes by confining attention to only two possible consequences: even development and uneven development. If the South prefers the process of even development to uneven development, then there is a welfare case for intervention if by doing so one takes the system from having uneven to even development. This kind of social welfare function, ranking only the two cases, has no relation to the standard welfare approaches which look at intertemporal choice functions over consumption bundles. From what is said in the text, one cannot defend intervention on such usual welfare grounds. Indeed, with shifting preference orderings, it would be difficult to compare the welfare implications of alternative paths.

27 See the previous chapter.

28 To show this, consider a variant of Findlay's (1980) model, where output per workers in the North and South, f and g, are functions of their capital-labour ratio k_i (that is, the production functions are the usual neoclassical kind), saving in the North is a fraction σ_n of total income, full employment growth takes place in the North (where labour supply grows at the fixed rate n), and the Southern real wage is fixed at V_s. The other assumptions and notation are the same as in our text. The Findlay model is thus a variant of the model with the neoclassical North and neo-Marxian South examined in the previous chapter.

The dynamics of this model can be analysed with the following equations of motion:

$$\hat{k}_n = \sigma_n f(k_n)/k_n - n$$

$$\hat{k} = f(k_n)/k_n \{\sigma_n - s_s(1-a)(1-\sigma_n)/[s_s + b(1-s_s)]\}$$

It is easy to check that the long-run equilibrium value of k is

$$k^* = [s_s + b(1-s_s)]\sigma_n/[(1-a)(1-\sigma_n)s_s]$$

and that the condition for uneven development (starting with $k > k^*$) is the same as that in our model. Several of our other conclusions also follow from this model as well.

29 The destabilizing effects of changes in expenditure patterns for development within an economy with a non-neoclassical model have been explored also by Pasinetti (1981).

30 See Von Weizsacker (1971).

10 Technological change and uneven development

1 See chapter 3 and Laibman (1982). Apart from the general arguments which can be made, we make this assumption to allow the real wage to rise due to technological

progress and terms of trade change, which is a crucial aspect of the story we are telling here. There are other possible assumptions we could have made instead: following Thirlwall (1986) W_n/P_s (the wage in terms of the Southern good, say food), could be fixed; so could W_n/P_n.

2 An implicit assumption behind this formulation is that technological progress in the two regions follows independent processes. We will consider interaction – due to technical competitiveness and technology transfer – in section 10.3.

3 In effect this makes technological progress exogenous in our model.

4 As a_0^s falls, the contribution of h_s obviously falls.

5 Findlay and Kierzkowski (1983) examine a model of trade with human capital formation. However, they allow such capital formation to result only from schooling and education, and not from the accumulation of experience. It should be emphasized that the experience we speak of is a concept far broader than on-the-job experience.

6 Even if the production of some goods in poor countries requires sophisticated techniques, this often occurs in enclaves which are integrated more closely with rich foreign countries than with the rest of the poor country, and thus have negligible spin-off effects.

7 As already noted, it is also non-capitalists in requiring no capital in production.

8 In the learning by doing model of chapter 5 the fruits of learning were enjoyed in the form of higher productivity in the sector where learning takes place (the sector there being the whole economy). The learning here can therefore be called generalized learning, a more complex involving also spin-off effects to other sectors.

9 Log-differentiation of these equations shows these equations are a special case and modification of Kaldor's technical progress functions, with the rate of growth of the capital in manufacturing to labour being relevant for the rate of growth of labour productivity in *both* sectors, and with the intercept term equal to zero (since our Θ_{ij} are constants). The dynamic infant industry models of Sheshinski (1967), Clemhout and Wan (1970), and Bardhan (1971), among others, are similar to ours in the way they model technological change. These models consider optimal policies in the presence of learning, where productivities depend on cumulative experience in some sense. However, they differ from our model in not allowing experience in one sector to have spin-off effects on other sectors, and in not considering the interaction between learning processes in two countries.

10 Formally, the model is a development and a modification of Krugman's (1981a) model. It is very similar to one of the models examined briefly in Dutt (1986a), but modifies it to make it conform more closely to the general framework used in chapter 8.

11 In terms of the notation that would be consistent with those of earlier chapters, $c = 1/a_1^j$, $b_{ij} = 1/a_{0i}^j$.

12 Note that since at long-run equilibrium we have capital and labour growing at the same rate, this implies that $n = a\sigma$, where σ is the saving-income ratio. Since in our model s is the saving-profit ratio, this requires $n < as$.

13 The assumption of incomplete specialization implies that in both countries X_a and X_m must be positive. From (10.15) and (10.20) this implies that

$$L - cK_j/\Theta_m k_j^{n_m} > 0.$$

This can be rewritten as

$$k_j < (\Theta_m/c)^{1/(1 - n_m)}$$

We assume this to be the case for all k_j we consider. In long-run equilibrium for the trading economy we will see that k_j is given by (10.27) which implies that

$$a(as-n)/(cs-\tau n)<1$$

which is satisfied since $a<1$ and $\tau<1$.

14 An example of the position that distortions are empirically unimportant is Haberler (1950). Since these distortions are the same factors which imply that a competitive economy does not attain a Pareto optimal state, the tendency for the apologists of unfettered private enterprise is understandable. The argument that with optimal corrective policies countries must gain by trading is logically unexceptionable. However, if these distortions become numerous and important, and in fact intrinsic aspects of the process of development, it becomes impossible for even the best intentioned governments to pursue such policies.

15 If neoclassical 'microfoundations' in terms of utility functions are desired, they can be provided along the lines pursued in Krugman (1979) and Dollar (1986).

11 Conclusion

1 Several of these extensions could involve the endogenization of some of the parameters of the different models, along the lines discussed in earlier chapters, especially chapter 3.

REFERENCES

Abraham-Frois, G. and Berrebi, E. (1979). *Theory of Value, Prices and Accumulation. A Mathematical Interpretation of Marx, Von Neumann and Sraffa*, Cambridge University Press.

Akerlof, George A. (1984). *An Economic Theorist's Book of Tales*, Cambridge University Press.

Amadeo, Edward (1986). 'Notes on capacity utilization, accumulation and distribution', *Contributions to Political Economy*, 5, March: 83–94.

(1989). *Keynes's Principle of Effective Demand*, Upleadon, Glos.: Edward Elgar.

Amin, Samir (1976). *Imperialism and Unequal Development*, New York: Monthly Review Press.

(1977). *Unequal Development*, New York: Monthly Review Press.

Arrow, Kenneth J. (1959). 'Toward a theory of price adjustment', in M. Abramovitz, *et al.*, eds., *The Allocation of Economic Resources*, Stanford: Stanford University Press.

(1962). 'The economic implications of learning by doing', *Review of Economic Studies*, 29(3), June: 155–73.

Asimakopulos, A. (1969). 'A Robinsonian growth model in one sector notation', *Australian Economic Papers*, 8, June: 41–58.

(1970). 'A Robinsonian growth model in one sector notation – an amendment', *Australian Economic Papers*, 9, December: 171–9.

Auerbach, Paul and Skott, Peter (1988). 'Concentration, competition and distribution – a critique of theories of monopoly capital', *International Review of Applied Economics*, 2.

Bacha, Edmar L. (1978). 'An interpretation of unequal exchange from Prebisch-Singer to Emmanuel', *Journal of Development Economics*, 5(4), December: 319–30.

Baran, Paul (1957). *The Political Economy of Growth*, New York: Monthly Review Press.

Baran, Paul and Sweezy, Paul (1966). *Monopoly Capitalism*, New York: Monthly Review Press.

Bardhan, Pranab (1970). *Economic Growth, Development and Foreign Trade*, New York: Wiley-Interscience.

(1971). 'On optimum subsidy to a learning industry: an aspect of the theory of infant industry protection', *International Economic Review*, 12(1), February: 54–70.

Bauer, Otto (1986). 'Otto Bauer's "Accumulation of Capital" (1913)', translated by J. E. King, *History of Political Economy*, 18(1), Spring: 87–110.

Baumol, William J. (1970). *Economic Dynamics*, 3rd. edn, New York: Macmillan.

Bhaduri, Amit (1986). *Macroeconomics – The Dynamics of Commodity Production*, Armonk, New York: M. E. Sharpe.

242

Bhaduri, Amit and Robinson, Joan (1980). 'Accumulation and exploitation: an analysis in the tradition of Marx, Sraffa and Kalecki', *Cambridge Journal of Economics*, 4(2), June: 103–15.

Bharadwaj, Krishna (1978). *Classical Political Economy and the Rise to Dominance of Supply and Demand Theories*, Culcutta: Orient Longman.

Black, J. (1962). 'The technical progress function and the production function', *Economica*, 29, May: 166–70.

Blatt, John M. (1983). *Dynamic Economic Systems. A Post-Keynesian Approach*, Armonk, New York: M. E. Sharpe.

Blaug, Mark (1960). 'Technical change and Marxian economics', *Kyklos*, 13(4): 495–512.

(1980). *The Methodology of Economists or How Economists Explain*, Cambridge University Press.

Bleaney, M. F. (1976). *Underconsumption Theories – A History and Critical Analysis*, New York: International Publishers.

Bliss, Christopher J. (1975). *Capital Theory and the Distribution of Income*, Amsterdam: North Holland.

Boggio, Luciano (1985). 'On the stability of production prices', *Metroeconomica*, 37(3), October: 241–67.

(1986). 'Stability of production prices in a model of general interdependence', in Semmler (1986).

Boland, Lawrence A. (1981). 'On the futility of criticizing the neoclassical maximization hypothesis', *American Economic Review*, 71(5), December: 1031–6.

(1985). 'The foundations of Keynes' methodology: *The General Theory*', in Tony Lawson and Hashem Pesaran, eds., *Keynes' Economics*, Armonk, New York: M. E. Sharpe.

Bowles, Samuel (1985). 'The production process in a competitive economy: Walrasian, neo-Hobbesian and Marxian models', *American Economic Review* 75(1), March: 16–36.

Bowles, Samuel and Boyer, Robert (1988). 'Labor discipline and aggregate demand: a macroeconomic model', *American Economic Review*, Papers and Proceedings, 78(2), May: 395–400.

Braun, Oscar (1984). *International Trade and Imperialism*, Atlantic Highlands, New Jersey: Humanities Press.

Brewer, Anthony (1980). *Marxist Theories of Imperialism – A Critical Survey*, London: Routledge and Kegan Paul.

(1985). 'Trade with fixed real wages and mobile capital', *Journal of International Economics*, 18(1/2), February: 177–86.

Brody, A. (1970). *Proportions, Prices and Planning*, Amsterdam: North Holland.

Bukharin, Nikolai (1915). *Imperialism and the World Economy*, London: Merlin, 1972.

Burgstaller, Andre (1985). 'North-South trade and capital flows in a Ricardian model of accumulation', *Journal of International Economics*, 18(3/4), May: 241–60.

Burgstaller, Andre and Saavedra-Rivano, Neantro (1984). 'Capital mobility and growth in a North–South model', *Journal of Development Economics*, 15(1/2/3), May–June–August: 213–37.

Burmeister, Edwin (1975). 'A comment on "This age of Leontieff... and who"', *Journal of Economic Literature*, 13(2), June: 454–7.

(1977). 'The irrelevance of Sraffa's analysis without returns to scale', *Journal of Economic Literature*, 15(1), March: 68–70.

(1980). *Capital Theory and Dynamics*, Cambridge University Press.

Caldwell, Bruce J. (1983). 'The neoclassical maximization hypothesis: comment', *American Economic Review*, 73(4), September: 824–7.

Cardoso, Eliana (1981). 'Food supply and inflation', *Journal of Development Economics*, 8(3), June: 269–84.

Chamberlin, Edward H. (1933). *The Theory of Monopolistic Competition*, Cambridge, Mass.: Harvard University Press.

Chang, W. W. and Smyth, D. J. (1971). 'The existence and persistence of cycles in a non-linear model: Kaldor's 1940 model re-examined', *Review of Economic Studies*, 38(1), January: 37–44.

Chichilnisky, Graciela (1981). 'Terms of trade and domestic distribution: export led growth with abundant labor', *Journal of Development Economics*, 8(2), April: 163–92.

(1984). 'North–South trade and export-led policies', *Journal of Development Economics*, 15(1/2/3), May–June–August: 131–60.

Chick, Victoria (1983). *Macroeconomics after Keynes*, Cambridge, Mass.: MIT Press.

Ciccone, R. (1986). 'Accumulation, utilisation of capacity and income distribution: some critical considerations on Joan Robinson's theory of distribution', *Political Economy*, 2(1):17–36.

Clemhout, S. and Wan, H. Y., Jr. (1970). 'Learning by doing and infant industry protection', *Review of Economic Studies*, 37(1), January: 33–56.

Clifton, James (1977). 'Competition and the evolution of the capitalist mode of production', *Cambridge Journal of Economics*, 1(2), June: 137–51.

(1983). 'Administered prices in the context of capitalist development', *Contributions to Political Economy*, 2, March: 23–38.

Coddington, Alan (1983). *Keynesian Economics – The Search for First Principles*, London: George Allen and Unwin.

Committeri, Marco (1986). 'Some comments on recent contributions on capital accumulation, income distribution and capacity utilization', *Political Economy*, 2(2):161–86.

Conway, Patrick and Darity, William A., Jr. (1985). 'Growth and trade with asymmetric returns to scale: a model for Nicholas Kaldor', mimeo, Department of Economics, University of North Carolina.

Cornwall, John (1977). *Modern Capitalism. Its Growth and Transformation*, New York: St Martin's Press.

Cowling, Kenneth (1982). *Monopoly Capitalism*, London: Macmillan.

Cowling, Kenneth and Waterson, M. (1976). 'Price cost margins and market structure', *Economica*, 43(171), August: 267–74.

Cross, Rod (1982). 'The Duhem-Quine thesis, Lakatos and the appraisal of theories in macroeconomics', *Economic Journal*, 92(366), June: 320–40.

Darity, William A., Jr. (1981). 'The simple analytics of neo-Ricardian growth and distribution', *American Economic Review*, 71(6), December: 978–93.

(1982a). 'On the long run outcome of the Lewis-Nurkse international growth process', *Journal of Development Economics*, 10(3), June: 271–8.

(1982b). 'A general equilibrium model of the eighteenth-century Atlantic slave trade', *Research in Economic History*, Volume 7:287–326, JAI Press.

(1987). 'Debt, finance, production and trade in a North–South model: the surplus approach', *Cambridge Journal of Economics*, 11(3), September: 211–27.

Davidson, Paul (1978). *Money and the Real World*, 2nd edn, London: Macmillan.

de Brunhoff, Susan (1976). *Marx on Money*, New York: Urizen.

Desai, Meghnad (1973). 'Growth cycles and inflation in a model of the class struggle', *Journal of Economic Theory*, 6, June: 527–45.

Dixit, Avinash (1977). 'The accumulation of capital theory', *Oxford Economic Papers*, 29(1), March: 1–29.

(1984). 'Growth and terms of trade under imperfect competition', in H. Kierzkowski, ed., *Monopolistic Competition and International Trade*, London: Oxford University Press

Dmitriev, V. (1974). *Economic Essays on Value, Competition and Utility*, in D. Nuti, ed., Cambridge University Press.

Dobb, Maurice (1940). *Political Economy and Capitalism*, London: Routledge.

Dollar, David (1986). 'Technological innovation, capital mobility, and the product cycle in North–South trade', *American Economic Review*, 76(1), March: 177–90.

Domar, Evsey D. (1957). 'A Soviet model of growth', in, E. Domar, *Essays in the Theory of Economic Growth*, New York: Oxford University Press.

Dornbusch, Rudiger and Fischer, Stanley (1978). *Macroeconomics*, New York: McGraw-Hill.

Dow, Sheila (1980). 'Methodological morality in the Cambridge controversies', *Journal of Post Keynesian Economics*, 2(3), Spring: 368–80.

(1985). *Macroeconomic Thought*, London: Basil Blackwell.

Drandakis, E. M. (1963). 'Factor substitution in the two-sector growth model', *Review of Economic Studies*, 30(3), October: 217–28.

Drandakis, E. M. and Phelps, E. S. (1966). 'A model of induced innovation, growth and distribution', *Economic Journal*, 76:823–40.

Duesenberry, James S. (1949). *Income, Saving and the Theory of Consumer Behaviour*, Cambridge, Mass.: Harvard University Press.

Dumenil, G. and Levy, D. (1986). 'Stability and instability in a dynamic model of capitalist reproduction', in Semmler (1986).

Dutt, Amitava K. (1982). 'Essays on growth and income distribution of an under-developed economy', unpublished Ph.D. dissertation, Massachusetts Institute of Technology.

(1984a). 'Stagnation, income distribution and monopoly power', *Cambridge Journal of Economics*, 8(1), March: 25–40.

(1984b). 'Uneven development and the terms of trade: a model of North–South trade with rigid prices', Discussion Paper No. 27, Florida International University, Sept.

(1986a). 'Vertical trading and uneven development', *Journal of Development Economics*, 20(2), March: 339–59.

(1986b). 'Stock equilibrium in fixprice – flexprice models for LDCs – the case of food speculation', *Journal of Development Economics*, 21(1), April: 89–109.

(1986c). 'Growth, distribution and technological change', *Metroeconomica*, 38(2), June: 113–34.

(1986–7). 'Wage rigidity and unemployment: the simple diagrammatics of two views', *Journal of Post Keynesian Economics*, 9(2), Winter: 279–90.

(1987a). 'Alternative closures again – a comment on "Growth, distribution and inflation"', *Cambridge Journal of Economics*, 11(1), March: 75–82.

(1987b) 'Keynes with a perfectly competitive goods market', *Australian Economic Papers*, 26(49), December: 275–93.

(1987c). 'The terms of trade and uneven development: implications of a model of North–South trade', *Pesquisa e Planejamento Economico*, 17(3), December: 533–60.

(1987d). 'Trade, debt and uneven development in a North–South model', unpublished, Florida International University, June.

(1988a). 'Monopoly power and uneven development: Baran revisited', *Journal of Development Studies*, 24(2), January: 161–76.

(1988b). 'Inelastic demand for Southern goods, international demonstration effects and uneven development', *Journal of Development Economics*, 29(1), July: 111–22.

(1988c). 'North–South models: a critical survey', unpublished, Department of Economics, University of Notre Dame.

(1988d). 'Convergence and equilibrium in two sector models of growth, distribution and prices', *Zeitschrift fur Nationalokonomie*, 48(2):135–58.

(1988e). 'Competition, monopoly power and the prices of production', *Thames Papers in Political Economy*, Autumn.

(1989a). 'Uneven development in alternative models of North–South trade', *Eastern Economic Journal*, forthcoming.

(1989b). 'Competition, monopoly power and the uniform rate of profit', *Review of Radical Political Economics*, forthcoming.

(1989c). 'Growth, distribution and capital ownership: Kalecki and Pasinetti revisited', unpublished, University of Notre Dame.

(1989d). 'Sectoral balance in development: a survey', *World Development*, forthcoming.

Eatwell, John (1977). 'The irrelevance of returns to scale in Sraffa's analysis', *Journal of Economic Literature*, 15(1), March: 61–8.

(1982). 'Competition' in Ian Bradley and Michael Howard, eds., *Classical and Marxian Political Economy*, New York: St. Martin's Press.

(1983). 'The long period theory of employment', *Cambridge Journal of Economics*, 7(3/4), September/December: 269–85.

Eatwell, John and Milgate, Murray, eds. (1983). *Keynes' Economics and the Theory of Value and Distribution*, Oxford and New York: Oxford University Press.

Eichner, Alfred S. (1976). *The Megacorp and Oligopoly*, Cambridge University Press.

Elster, Jon (1979). *Ulysses and the Sirens, Studies in Rationality and Irrationality*, Cambridge University Press, Paris: Editions de la Maison des Sciences de L'Homme.

(1983). *Explaining Technical Change*, Cambridge University Press.

Eltis, W. A. (1973). *Growth and Distribution*, London: Macmillan.

Emmanuel, Arghiri (1972). *Unequal Exchange: A study of the Imperialism of Trade*, New York: Monthly Review Press.

(1978). 'A note on trade pattern reversals', *Journal of International Economics*, 8(1), February: 143–5.

Evans, David (1976). 'Unequal exchange and economic policies: some implications of the neo-Ricardian critique of the theory of comparative advantage', *Economic and Political Weekly*, 11(5–7), February.

Feiwell, G. R. (1975). *The Intellectual Capital of Michal Kalecki*, Knoxville: University of Tennessee Press.

Feyerabend, Paul (1975). *Against Method*, London: New Left Books.

Findlay, Ronald (1963). 'The Robinsonian model of accumulation', *Economica*, 30(117): 1–12.

(1980). 'The terms of trade and equilibrium growth in the world economy', *American Economic Review*, 70(3), June: 291–9.

(1981). 'Fundamental determinants of the terms of trade', in S. Grassman and E. Lundberg, eds., *The World Economic Order – Past and Prospects*, London: Macmillan.

Findlay, Ronald and Kierzkowski, H. (1983). 'International trade and human capital: a simple general equilibrium model', *Journal of Political Economy*, 91(6), December: 957–78.

Fisher, Franklin (1983). *Disequilibrium Foundations of Equilibrium Economics*, Cambridge University Press.

Flanders, M. J. (1964). 'Prebisch on protectionism: an evaluation', *Economic Journal*, 74(294), June: 305–26.

Flaschel, Peter and Semmler, Willi (1986a). 'Classical and neoclassical adjustment processes', *Manchester School*, 55(1), March: 13–37.

(1986b). 'The dynamic equalization of profit rates for input-output models with fixed capital', in Semmler (1986).

Foley, Duncan (1983). 'On Marx's theory of money', *Social Concept*, May.

(1986). *Money, Accumulation and Crisis*, Chur: Harwood.

Foster, J. and Szlajfer, H., eds. (1984). *The Faltering Economy*, New York: Monthly Review Press.

Frank, Andre Gunder (1975). *On Capitalist Underdevelopment*, Oxford University Press.

Franke, Reiner (1986). 'A cross-over gravitation process in prices and inventories', in Semmler (1986).

Fujimoto, Takao and Krause, Ulrich (1986). 'Ergodic price setting with technical progress', in Semmler (1986).

Galtung, Johan (1971). 'A structural theory of imperialism', *Journal of Peace Research*, 8(2): 81–117.

Gandolfo, Giancarlo (1980). *Economic Dynamics: Methods and Models*, Amsterdam: North Holland.

Garegnani, Pierangelo (1976). 'On a change in the notion of equilibrium in recent work on value and distribution', in M. Brown, K. Sato, and P. Zarembka, eds., *Essays in Modern Capital Theory*, Amsterdam: North Holland, repr. in Eatwell and Milgate (1983).

(1978–79). 'Notes on consumption, investment and effective demand', Parts I and II, *Cambridge Journal of Economics*, Dec. 1978, Mar. 1979, repr. in Eatwell and Milgate (1983).

(1983). 'The classical theory of wages and the role of demand schedules in the determination of relative prices', *American Economic Review*, Papers and Proceedings, 73(2), May: 309–13.

(1984). 'Value and distribution in the classical economists and Marx', *Oxford Economic Papers*, 36(2), June: 291–325.

Gerschenkron, Alexander (1962). *Economic Backwardness in Historical Perspective*, Cambridge, Mass.: Harvard University Press.

Gibson, Bill (1980). 'Unequal exchange: theoretical issues and empirical findings', *The Review of Radical Political Economics*, 12(3), Fall: 15–35.

Glick, Mark (1985). 'Monopoly or competition in the US economy?', *Review of Radical Political Economics*, 17(4), Winter: 121–27.

Glombowski, Jorg (1983). 'A Marxian model of long run capitalist development', *Zeitschrift fur Nationalokonomie*, 43(4): 363–82.

(1986). 'Comment' in H.-J. Wagener and J. W. Drukker, eds., *The Economic Law of Motion of Modern Society – A Marx–Keynes–Schumpeter Centennial*, Cambridge University Press.

Goldstein, Jonathan P. (1985). 'Pricing, accumulation, and crisis in the Keynesian theory', *Journal of Post Keynesian Economics*, 8(1), Fall: 121–34.

Goodwin, Richard M. (1951). 'The non-linear accelerator and the persistence of business cycles', *Econometrica*, 19: 1–17, repr. in Goodwin (1982).

(1955). 'A model of cyclical growth' in E. Lundberg, ed., *The Business Cycle in the Post-War World*, London: Macmillan, repr. in Goodwin (1982).

(1967). 'A growth cycle', in C. H. Feinstein, ed., *Socialism, Capitalism and Growth*, Cambridge University Press, repr. in Goodwin (1982).

(1982). *Essays in Economic Dynamics*, London: Macmillan.

Goodwin, Richard M., Kruger, M. and Vercelli, A. (1984). *Nonlinear Models of Fluctuating Growth*, Berlin: Springer-Verlag.

Haberler, Gottfried (1950). 'Some problems in the pure theory of international trade', *Economic Journal*, 60, June: 223–40.

(1961). 'Terms of trade and economic development', in H. Ellis, ed., *Economic Development in Latin America*, New York: St Martin's Press.

Hacche, Graham (1979). *The Theory of Economic Growth – An Introduction*, New York: St Martin's Press.

Hahn, Frank (1969). 'On money and growth', *Journal of Money, Credit and Banking*, 1(2), May: 172–87.

(1975). 'Revival of political economy: the wrong issues and the wrong argument', *Economic Record*, 51(135), September: 360–4.

(1982). 'The neo-Ricardians', *Cambridge Journal of Economics*, 6(4), December: 353–74.

(1986). 'Of Marx and Keynes and many things', *Oxford Economic Papers*, 38(2), July: 354–61.

Hamilton, Alexander (1791). 'Report on manufactures', repr. in Jacob E. Cooke, eds., *The Reports of Alexander Hamilton*, New York: Harper and Row.

Harcourt, Geoffrey C. (1963). 'A critique of Mr. Kaldor's model of income distribution and economic growth', *Australian Economic Papers*, 2(1), June, repr. in Harcourt (1982).

(1965). 'A two-sector model of the distribution of income and the level of employment in the short run', *Economic Record*, 41(93), March, repr. in Harcourt (1982).

(1972). *Some Cambridge Controversies in the Theory of Capital*, Cambridge University Press.

(1981). 'Marshall, Sraffa and Keynes: incompatible bedfellows?', *Eastern Economic Journal*, 7(1), repr. in Harcourt (1982).

(1982). *The Social Science Imperialists*, London: Routledge and Kegan Paul.

Harcourt, Geoffrey C. and Kenyon, Peter (1976). 'Pricing and the investment decision', *Kyklos*, 29(3): 449–77.

Harcourt, Geoffrey C. and Massaro, Vincent G. (1964). 'Mr. Sraffa's production of commodities', *Economic Record*, 40(91), September, repr. (amended) in Harcourt (1982).

Harris, Donald J. (1978). *Capital Accumulation and Income Distribution*, Stanford, California: Stanford University Press.

(1981). 'On the timing of wage payments', *Cambridge Journal of Economics*, 5(4), December: 369–81.

(1983). 'Accumulation and the rate of profit in Marxian theory', *Cambridge Journal of Economics*, 7(3/4), September–December: 311–30.

(1985). 'The theory of economic growth: from steady states to uneven development', in G. R. Feiwel, ed., *Issues in Contemporary Macroeconomics and Distribution*, Albany: State University of New York Press.

(1986). 'Are there macroeconomic laws? The "law" of the falling rate of profit reconsidered', in H.-J. Wagener and H. W. Drukker, eds. *The Economic Law of Motion of Modern Society – A Marx–Keynes–Schumpeter Centennial*, Cambridge University Press.

(1988). 'On the classical theory of competition', *Cambridge Journal of Economics*, 12(1), March: 139–67.

Harrod, Roy F. (1936). *The Trade Cycle*, Oxford University Press.

(1939). 'An essay in dynamic theory', *Economic Journal*, 49: 14–33.

(1948). *Towards a Dynamic Economics*, London: Macmillan.

(1973). *Economic Dynamics*, London: Macmillan.

Hausman, Daniel M. (1981). *Capital, Profits, and Prices. An Enquiry in the Philosophy of Economics*, New York: Columbia University Press.

Heertje, Arnold (1977). *Economics and Technological Change*, New York: John Wiley and Sons.

Hicks, John R. (1932). *The Theory of Wages*, London: Macmillan. 2nd ed., 1963.

(1949). 'Mr. Harrod's Dynamic Theory', *Economica*, 16: 106–21.

(1950). *A Contribution to the Theory of the Trade Cycle*, London: Oxford University Press.

(1965). *Capital and Growth*, London: Oxford University Press.

(1974). *The Crisis in Keynesian Economics*, London: Oxford University Press.

(1979). *Causality in Economics*, New York: Basic Books.

(1985). *Methods of Dynamic Economics*, Oxford University Press.

Hilferding, Rudolf (1910). *Finance Capital*, English trans., London and Boston: Routledge and Kegan Paul, 1981.

Hollander, Samuel (1979). *The Economics of David Ricardo*, Toronto: University of Toronto Press.

(1985). 'On the substantive identity of the Ricardian and neoclassical conceptions of economic organization: the French connection in British classicism', in Carvale, ed. (1985).

Hosoda, Eiji (1985). 'On the classical convergence theorem', *Metroeconomica*, 36(2), June: 157–74.

Inada, K. (1963). 'On a two-sector model of economic growth: comments and a generalization', *Review of Economic Studies*, 30(2) No. 83, June: 119–27.

Johansen, Leif (1960). *A Multisectoral Study of Economic Growth*, Amsterdam: North Holland.

Jones, Ronald W. (1965). 'The structure of simple general equilibrium models', *Journal of Political Economy*, 73(6), December: 557–72.

(1971). 'A three-factor model in theory, trade and history', in J. N. Bhagwati, R. W. Jones, R. A. Mundell, and J. Vanek, eds., *Trade, Balance of Payments and Growth*, Amsterdam: North-Holland.

Kaldor, Nicholas (1940). 'A model of the trade cycle', *Economic Journal* 50: 78–92, repr. in Kaldor (1960).

(1955–6). 'Alternative theories of distribution', *Review of Economics Studies*, 23(2), No. 61: 83–100.

(1957). 'A model of economic growth', *Economic Journal*, 67: 591–624, repr. in Kaldor (1960).

(1959). 'Economic growth and the problem of inflation', *Economica*, 26: 212–26, 287–98.

(1960). *Essays on Economic Stability and Growth*, London: Duckworth.

(1961). 'Capital accumulation and economic growth', in F. A. Lutz and D. C. Hague, eds., *The Theory of Capital Accumulation*, London: Macmillan.

(1966). *Causes of the Slow Rate of Economic Growth of the United Kingdom*, Cambridge University Press.

(1979). 'Equilibrium theory and growth theory', in M. J. Boskin, ed., *Economics and Human Welfare*, London and New York: Academic Press.

(1982). *The Scourge of Monetarism*, London: Oxford University Press.

(1986). 'Limits on growth', *Oxford Economic Papers*, 38(2), July: 187–98.

Kalecki, Michal (1933). 'Outline of the theory of the business cycle', in *An Essay on the Theory of the Business Cycle*, repr. in Kalecki (1971).

(1935). 'A macrodynamic theory of business cycles', *Econometrica*, 3: 327–44.

(1939). *Essays in the Theory of Economic Fluctuations*, London: Allen and Unwin.

(1939–40). 'The supply curve of an industry under imperfect competition', *Review of Economic Studies*, 7: 91–112.

(1941). 'A theorem on technical progress', *Review of Economic Studies*, 8.

(1941a). 'A theory of long-run distribution of the product of industry', *Oxford Economic Papers*, NS, No. 6, pp. 31–41.

(1942). 'Mr. Whitman on the concept of "Degree of Monopoly"', *Economic Journal*, 52, April: 121–7.

(1943). *Studies in Economic Dynamics*, London: Allen and Unwin.

(1954). *Theory of Economic Dynamics*, London: Allen and Unwin.

(1962). 'A model of hyperinflation', *Manchester School*, 30(3), September: 275–82.

(1968). 'Trend and the business cycle', *Economic Journal*, 78, repr. in Kalecki (1971).

(1971). *Selected Essays on the Dynamics of the Capitalist Economy, 1933–1970*, Cambridge University Press.

(1971a). 'Class struggle and distribution of national income', *Kyklos*, 24: 1–9, repr. in Kalecki (1971).

Kennedy, C. (1964). 'Induced bias in innovation and the theory of distribution', *Economics*, 4(1), March: 23–36, repr. in Eatwell and Milgate, eds. (1983).

Keynes, John Maynard (1930). *A Treatise on Money*, London: Macmillan.

(1936). *The General Theory of Employment, Interest and Money*, London: Macmillan.

Kindleberger, Charles P. (1956). *The Terms of Trade – an European Case Study*, New York: John Wiley and Sons.

Kindleberger, Charles P. (1956). *The terms of trade – an European case study*, New York: John Wiley and Sons.

Kottis, G. (1971). 'The international demonstration effect as a factor affecting development', *Kyklos*, 74: 455–72.

Kotz, David M. (1982). 'Monopoly, inflation and economic crisis', *Review of Radical Political Economics*, 14(4), Winter: 1–17.

Kriesler, Peter (1987). *Kalecki's Microanalysis*, Cambridge University Press.

Krugman, Paul (1979). 'A model of innovation, technology transfer, and the world distribution of income', *Journal of Political Economy*, 87(2), April: 253–66.

(1981a). 'Trade, accumulation and uneven development', *Journal of Development Economics*, 8(2), April: 149–62.

(1981b). 'Intraindustry specialization and the gains from trade', *Journal of Political Economy*, 89(5), October: 959–73.

Kuroki, Ryuzo (1986). 'The equalization of the rate of profit reconsidered', in Semmler (1986).

Kurz, Heinz D. (1985). 'Sraffa's contribution to the debate in capital theory', *Contributions to Political Economy*, 4, March: 3–24.

(1986). '"Normal" positions and capital utilisation', *Political Economy*, 2(1): 37–54.

Laibman, David (1982). 'Technical change, the real wage and the rate of exploitation. The falling rate of profit reconsidered', *Review of Radical Political Economics*, 4(2), Summer: 95–105.

Lavoie, Don (1986). 'Marx, the quantity theory, and the theory of value', *History of Political Economy*, 18(1), Spring: 155–70.

Lazonick, William (1979). 'Industrial relations and technical change: the case of the self-acting mule', *Cambridge Journal of Economics*, 3(3), September: 231–62.

Leijonhufvud, Axel (1968). *On Keynesian Economics and the Economics of Keynes*, Oxford University Press.

(1974). 'Keynes's employment function', *History of Political Economy*, 6(2), Summer: 164–70.

Lenin, Vladimir I. (1917). *Imperialism, the Highest Stage of Capitalism*, Moscow: Foreign Languages Publishing House, 1950.

Levhari, David (1966). 'Extensions of Arrow's learning by doing', *Review of Economic Studies*, 33, April: 117–31.

Levine, A. L. (1974). 'This age of Leontief ... and who? an interpretation', *Journal of Economic Literature*, 12(3), September: 872–81.

(1975). ' "This age of Leontief ... and who?" a reply', *Journal of Economic Literature*, 13(2), June: 457–81.

(1977). 'The irrelevance of returns to scale in Sraffa's analysis: a comment', *Journal of Economic Literature*, 15(1), March: 70–2.

(1985). 'Sraffa's *Production of Commodities by Means of Commodities*, returns to scale, relevance, and other matters', *Journal of Post Keynesian Economics*, 7(3), Spring: 342–9.

Levine, David P. (1980). 'Aspects of the classical theory of markets', *Australian Economic Papers*, 19(34), June: 1–15.

Lewis, W. Arthur (1954). 'Economic development with unlimited supplies of labour', *Manchester School*, 22(2), May: 139–91.

(1969). *Aspects of Tropical Trade 1883–1965*, Stockholm: Almqvist and Wicksell.

(1978). *The Evolution of the International Economic Order*, Princeton: Princeton University Press.

List, Friedrich (1841). *The National System of Political Economy* (original German; translation, New York: Longmans, Green and Co., 1904).

Mahalanobis, Prasanta C. (1953). 'Some observations on the process of growth of national income', *Sankhya*, 12: 307–12.

Mainwaring, Lynn (1977). 'Monopoly power, income distribution and price determination', *Kyklos*, 30(4): 674–90.

(1979). 'The wage-profit relation without constant returns', *Metroeconomica*, 30(3), October: 335–48.

(1980). 'International investment and the Pasinetti process', *Oxford Economic Papers*, 32(1), March: 99–101.

(1984). *Value and Distribution in Capitalist Economies*, Cambridge University Press.

Maneschi, Andrea (1986). 'A comparative evaluation of Sraffa's "The laws of returns under competitive conditions" and its Italian precursor', *Cambridge Journal of Economics*, 10(1), March: 1–12.

Marglin, Stephen A. (1974). 'What do bosses do? Part I', *Review of Radical Political Economics*, 6(2), Summer: 60–112.

(1984a). *Growth, Distribution and Prices*, Cambridge, Mass.: Harvard University Press.

(1984b). 'Growth, distribution and inflation – a centennial synthesis', *Cambridge Journal of Economics*, 8(2), June: 115–44.

Marglin, Stephen A. and Bhaduri, Amit (1986). 'Distribution, capacity utilization and growth', mimeo, March, Havard University.

Marshall, Alfred (1890). *Principles of Economics*, London: Macmillan (8th edition, 1920).

Marx, Karl (1857–8). *Grundrisse*, Harmondsworth: Penguin, 1973.

(1867). *Capital*, Vol. I, New York: International Publishers, 1967.

(1885). *Capital*, Vol. II, New York: International Publishers, 1967.

(1894). *Capital*, Vol. III, New York: International Publishers, 1967.

Michaely, Michael (1984). *Trade, Income Levels, and Dependence*, Amsterdam: North-Holland.

Milgate, Murray (1979). 'On the origin of the notion of "intertemporal equilibrium"', *Economica*, 46(181), February: 1–10.

(1982). *Capital and Employment – a Study of Keynes's Economics*, London and New York: Academic Press.

Moore, Basil J. (1988). *Horizontalists and Verticalists. The Macroeconomics of Credit Money*, Cambridge University Press.

Morishima, Michio (1973). *Marx's Economics. A Dual Theory of Value and Growth*, Cambridge University Press.

Morishima, Michio and Catephores, G. (1978). *Value, Exploitation and Growth*, New York: McGraw-Hill.

Mukherji, Badal (1982). *The Theory of Growth and the Tradition of Ricardian Dynamics*, Delhi: Oxford University Press.

Myrdal, Gunnar (1956). *An International Economy: Problems and Prospects*, London: Routledge and Kegan Paul.

(1957). *Economic Theory and Underdeveloped Regions*, London: Duckworth and Co.

Negishi, Takashi (1985). *Economic Theories in a Non-Walrasian Tradition*, Cambridge University Press.

Nell, Edward J. (1985). 'Jean Baptiste Marglin: a comment on "Growth, Distribution and Inflation"', *Cambridge Journal of Economics*, 9(2), June: 173–8.

Nelson, Richard R, and Winter, Sidney G. (1982). *The Evolutionary Theory of Economic Change*, Cambridge, Mass.: Harvard University Press.

Nikaido, Hokukane (1983). 'Marx on competition', *Zeitschrift fur Nationalokonomie*, 43(4): 337–62.

(1985). 'Dynamics of growth and capital mobility in Marx's scheme of reproduction', *Zeitschrift fur Nationalokonomie*, 45(3): 197–218.

Nurkse, Ragnar (1953). *Problems of Capital Formation in Under-developed Countries*, London: Oxford University Press.

Okishio, Nobuo (1961). 'Technical change and the rate of profit', *Kobe University Economic Review*, 7: 85–99.

(1977). 'Notes on technical progess and capitalist society', *Cambridge Journal of Economics*, 1(1), March: 93–100.

Panico, Carlo (1980). 'Marx's analysis of the relationship between the rate of interest and the rate of profit', *Cambridge Journal of Economics*, 4(4), December: 363–78, repr. in Eatwell and Milgate, eds. (1983).

(1985). 'Market forces and the relation between the rates of interest and profits', *Contributions to Political Economy*, 4, March: 36–60.

Pasinetti, Luigi (1962). 'The rate of profit and income distribution in relation to the rate of economic growth', *Review of Economic Studies*, 29: 267–79. Repr. in Pasinetti, L., *Growth and Income Distribution: Essays in Economic Theory*, Cambridge University Press, 1974.

(1977). *Lectures on the Theory of Production*, London: Macmillan.

(1981). *Structural Change and Economic Growth*, Cambridge University Press.

Patinkin, Don (1965). *Money, Interest and Prices*, 2nd ed., New York and London: Harper & Row.

Phelps, E. S. (1961). 'The golden rule of accumulation: a fable for growthmen', *American Economic Review*, 51(4), September: 638–43.

Pigou, A. C. (1943). 'The classical stationary state', *Economic Journal*, 53, December: 343–51.

Pivetti, Massimo (1985). 'On the monetary explanation of distribution', *Political Economy*, 1(2): 73–103.

Ploeg, Frederick van der (1983). 'Predator-prey and neoclassical models of cyclical growth', *Zeitschrift fur Nationalokonomie*, 43(3), 235–356.

Prebisch, Raul (1950). *The Economic Development of Latin America and its Principal Problems*, New York: ECLA.

(1959). 'Commercial policy in underdeveloped countries', *American Economic Review*, Papers and Proceedings, 49(2), May: 251–73.

(1963) *Towards a Dynamic Development Policy for Latin America*, New York: United Nations.

Pulling, K. (1978). 'Cyclical behaviour of profit margins', *Journal of Economic Issues*, 12(2), June: 287–305.

Radnitzky, Gerard and Bernholz, Peter, eds. (1987). *Economic Imperialism*, New York: Paragon House.

Ricardo, David (1817). *Principles of Political Economy and Taxation*, ed. P. Sraffa, Cambridge University Press, 1951.

Robinson, Joan V. (1933). *The Economics of Imperfect Competition*, London: Macmillan.

(1942). *An Essay on Marxian Economics*, London: Macmillan.

(1953). 'The production function and the theory of capital', *Review of Economic Studies*, 21: 81–106.

(1956). *The Accumulation of Capital*, London: Macmillan.

(1962). *Essays in the Theory of Economic Growth*, London: Macmillan.

Roche, John (1985). 'Marx's theory of money: a reinterpretation', *Review of Radical Political Economics*, 17(1/2), Summer: 201–11.

Roemer, John E. (1977). 'Technical change and the "tendency of the rate of profit to fall"', *Journal of Economic Theory*, 16(2), December: 403–24.

(1978). 'Marxian models of reproduction and accumulation', *Cambridge Journal of Economics*, 2(1), March: 37–53.

(1979). 'Continuing controversy on the falling rate of profit: fixed capital and other issues', *Cambridge Journal of Economics*, 3(4), December: 379–98.

(1981). *Analytical Foundations of Marxian Economic Theory*, Cambridge University Press.

(1982). *A General Theory of Exploitation and Class*, Cambridge, Mass.: Harvard University Press.

Roncaglia, Alessandro (1978). *Sraffa and the Theory of Prices*, Chichester: Wiley.

(1982). 'Hollander's Ricardo', *Journal of Post Keynesian Economics*, 4(3), Spring: 339–59.

Rosdolsky, Roman (1977). *The Making of Marx's 'Capital'*, London: Pluto Press.

Rotemberg, Julio J. and Saloner, Garth (1986). 'A supergame-theoretic model of price wars during booms', *American Economic Review*, 76(3), June: 390–407.

Rousseas, Stephen (1986). *Post Keynesian Monetary Economics*, Armonk, New York: M. E. Sharpe.

Rowthorn, Bob (1977). 'Conflict, inflation and money', *Cambridge Journal of Economics*, 1(3), September: 215–39, repr. in Rowthorn (1980).

(1980). *Capitalism, Conflict and Inflation*, London: Lawrence and Wishart.

(1982). 'Demand, real wages and economic growth', *Studi Economici*, 18.

Samuelson, Paul A. (1939). 'Interactions between the multiplier analysis and the principle of acceleration', *Review of Economics and Statistics*, 21: 75–8.

(1965). 'The theory of induced innovation along Kennedy-Weizsacker lines', *Review of Economics and Statistics*, 47, November: 343–56.

(1966). 'Rejoinder: agreements, disagreements, doubts and the case of induced Harrod-neutral technical change', *Review of Economics and Statistics*, 48:444–8.

(1973). 'Deadweight loss in international trade from the profit motive', in C. F. Bernstein and W. G. Tyler, eds., *Leading Issues in International Economic Policy*, Lexington, Mass.: D. C. Heath and Co.

(1975). 'Trade pattern reversals in time-phased Ricardian systems and intertemporal efficiency', *Journal of International Economics*, 5(4), November: 309–64.

(1976). 'Illogic of neo-Marxian doctrine of unequal exchange', in D. A. Belsey, E. J. Kane, P. A. Samuelson, and R. M. Solow, eds., *Inflation, Trade and Taxes, Essays in Honor of Alice Bourneuff*, Columbus: Ohio State University Press.

(1978). 'Free trade's intertemporal Pareto-optimality', *Journal of International Economics*, 8(1), February: 147–9.

Sardoni, Claudio (1981). 'Multi-sectoral models of balanced growth and the Marxian schemes of expanded reproduction', *Australian Economic Papers*, 20(37), December: 383–97.

(1986). 'Marx and Keynes on effective demand and unemployment', *History of Political Economy*, 18(3), Fall: 419–41.

Sarkar, Prabirjit (1986). 'The Singer-Prebisch hypothesis: a statistical evaluation', *Cambridge Journal of Economics*, 10(4), December: 355–71.

Sawyer, Malcolm C. (1982). *Macroeconomics in Question*, Armonk, New York: M.C. Sharpe.

(1985). *The Economics of Michal Kalecki*, London: Macmillan.

Schefold, Bertram (1980). 'Von Neumann and Sraffa: mathematical equivalence and conceptual difference', *Economic Journal*, 90(357): 140–56.

Scherer, F. M. (1980). *Industrial Market Structure and Economic Performance*, 2nd ed., Chicago: Rand McNally.

Schlicht, Ekkehart (1985). *Isolation and Aggregation in Economics*, Berlin: Springer-Verlag.

Schumpeter, Joseph (1934). *Theory of Capitalist Development*, Cambridge, Mass.: Harvard University Press.

Schwartz, Jacob (1961). *Lectures on the Mathematical Method in Analytical Economics*, New York: Gordon and Breach.

Semmler, Willi (1982). 'Competition, monopoly, and differential profit rates: theoretical considerations and empirical evidence', *Review of Radical Political Economics*, 13(4), Winter: 39–52.

(1984). *Competition, Monopoly and Differential Profit Rates*, New York: Columbia University Press.

(1986). ed., *Competition, Instability and Nonlinear Cycles*, Berlin: Springer-Verlag.

Sen, Amartya K. (1963). 'Neo-Classical and neo-Keynesian theories of distribution', *Economic Record*, 39: 53–64.

(1978). 'On the labour theory of value: some methodological issues', *Cambridge Journal of Economics*, 2(2), June: 175–90.

Shackle, G. L. S. (1984). 'General thought-schemes and the economist', *Thames Papers in Political Economy*, Autumn.

Shah, A. and Desai, M. (1981). 'Growth cycles with induced technical change', *Economic Journal*, 91(364), December: 1006–10.

Shaikh, Anwar (1978). 'Political economy and capitalism: notes on Dobb's theory of crisis', *Cambridge Journal of Economics*, 2(2), June: 233–51.

(1980). 'Marxian competition versus perfect competition: further comments on the so-called choice of technique', *Cambridge Journal of Economics*, 4(1), March: 75–83.

(1982). 'Neo-Ricardian economics: a wealth of algebra, a poverty of theory', *Review of Radical Political Economics*, 14(2), Summer: 67–83.

Shapiro, Nina (1978). 'Keynes and equilibrium economics', *Australian Economic Papers*, 17(31), December: 207–23.

Sherman, Howard J. (1968). *Profits in the United States*, Ithaca: Cornell University Press.

(1983). 'Monopoly power and profit rates', *Review of Radical Political Economics*, 15(2), Summer: 225–32.

(1987). 'The business cycle of capitalism', *International Review of Applied Economics*, 1(1): 72–85.

Sheshinski, Eytan (1967). 'Optimal accumulation with learning by doing', in Karl Shell, ed., *Essays on the Theory of Optimal Economic Growth*, Cambridge, Mass.: MIT Press.

Simon, Herbert A. (1979). 'Rational decision-making in business organisation', *American Economic Review*, 69(4), September: 493–513.

Singer, Hans (1950). 'The distribution of gains between investing and borrowing countries', *American Economic Review*, Papers and Proceedings, 40(2), May: 473–85.

(1975). *The Strategy of International Development*, International Arts and Science Press.

Skott, Peter (1989). 'Effective demand, class struggle and cyclical growth', *International Economic Review*, 30(1), February: 231–47.

Smith, Adam (1776). *An Inquiry into the Nature and Causes of the Wealth of Nations*, 2 Vols., New York and London: Oxford University Press, 1978.

Sodersten, Bo (1971). *International Economics*, London: Macmillan.

Solow, Robert M. (1956). 'A contribution to the theory of economic growth', *Quarterly Journal of Economics*, 70:65–94.

(1957). 'Technical change and the aggregate production function', *Review of Economics and Statistics*, 39(3), August: 312–20.

(1961). 'A note on Uzawa's two-sector model of economic growth', *Review of Economic Studies*, 29:48–50.

(1979). 'Another possible source of wage stickiness', *Journal of Macroeconomics*, 1(1), Winter: 79–82.

(1985). 'Economic history and economics', *American Economic Review*, Papers and Proceedings, 75(2), May: 328–31.

Spraos, John (1983). *Inequalising trade?* New York: Oxford University Press.

Sraffa, Piero (1960). *Production of Commodities by Means of Commodities*, Cambridge University Press.

Steedman, Ian (1977). *Marx After Sraffa*, London: New Left Books.

(1984). 'Natural prices, differential profit rates and the classical competitive process', *Manchester School*, 52(2), June: 123–40.

Steedman, Ian, Sweezy, Paul, *et al.* (1981). *The Value Controversy*, London: New Left Books.

Stein, Jerome L. (1969). '"Neoclassical" and "Keynes-Wicksell" monetary growth models', *Journal of Money, Credit and Banking*, 1(2), May: 123–40.

(1971). *Money and Capacity Growth*, New York and London: Columbia University Press.

Steindl, Josef (1952). *Maturity and Stagnation in American Capitalism*, Oxford: Basil Blackwell.

Stigler, George and Becker, Gary (1977). 'De gustibus non est disputandum', *American Economic Review*, 67(2), March: 76–90.

Stiglitz, Joseph E. (1984). 'Price rigidities and market structure', *American Economic Review*, Papers and Proceedings, 74(2), May: 350–5.

Sweezy, Paul (1942). *The Theory of Capitalist Development*, New York: Monthly Review Press.

Sylos-Labini, Paolo (1956). *Oligopoly and Technical Progress*, Cambridge, Mass.: Harvard University Press.

Taylor, Lance (1979). *Macro Models for Developing Countries*, New York: McGraw-Hill.

(1981). 'South-North trade and Southern growth: bleak prospects from a structuralist point of view', *Journal of International Economics*, 11(4), November: 589–602.

(1982). 'Food price inflation, the terms of trade and growth', in M. Gersovitz, C. F. Diaz-Alejandro, G. Ranis, and M. Rosenzwieg, eds., *The Theory and Experience of Economic Development*, London: George Allen and Unwin.

(1983). *Structuralist Macroeconomics*, New York: Basic Books.

(1984). 'Demand composition, income distribution and growth', mimeo, Massachusetts Institute of Technology.

(1985). 'A stagnationist model of economic growth', *Cambridge Journal of Economics*.

(1986). 'Debt crisis. North-South, North-North and in between', in M. P. Clauden, ed., *World Debt Crisis: International Lending on Trials*, Cambridge, Mass.: Ballinger.

Thirlwall, A. P. (1986). 'A general model of growth and development on Kaldorian lines', *Oxford Economic Papers*, 38(2), July: 199–219.

Tobin, James (1955). 'A dynamic aggregative model', *Journal of Political Economy*, 63(2), April: 103–15.

(1965). 'Money and economic growth', *Econometrica*, 33:671–84.

Uzawa, Hirofuni (1961). 'On a two-sector model of economic growth', *Review of Economic Studies*, 29(1), No. 78, October: 40–7.

(1963). 'On a two-sector model of economic growth, II', *Review of Economic Studies*, 30(2), No. 83, June: 105–18.

(1965). 'Optimum technical change in an aggregate model of economic growth', *International Economic Review*, 6, January: 18–31.

Van Parijs, Philippe (1980). 'The falling-rate-of-profit theory of crisis: a rational reconstruction by way of an obituary', *Review of Radical Political Economics*, 12(1), Spring: 1–17.

Varga, Eugene (1948). 'Changes in the economy of capitalism resulting from the Second World War', mimeo, Washington.

Veblen, Thorstein (1904). *The Theory of Business Enterprise*, New York, Scribner's.

(1915). *Imperial Germany and the Industrial Revolution*, London: Macmillan.

(1920). *Vested Interests and the Common Man*, New York, Huebsch.

(1923). *Absentee Ownership and Business Enterprise in Recent Times: the Case of America*, New York, Huebsch.

Vianello, Fernando (1985). 'The pace of accumulation', *Political Economy*, 1(1): 69–87.

Vines, David (1984). 'A North-South growth model along Kaldorian lines', Discussion Paper Series, No. 26, Centre for Economic Policy Research, Cambridge.

Von Bortkiewicz, L. (1906–7). 'Value and price in the Marxian system', Eng. tr., *International Economic Papers*, No. 2, 1952.

Von Neumann, J. (1945–6). 'A model of general economic equilibrium', *Review of Economic Studies*, 13: 1–9.

Von Weizsacker, C. C. (1971). 'Notes on endogenous change of tastes', *Journal of Economic Theory*, 3(4), December: 345–72.

Wallerstein, Immanuel (1974). *The Modern World System*, New York: Academic Press.

Weeks, John (1981). *Capital and Exploitation*, Princeton, New Jersey: Princeton University Press.

Weintraub, E. Roy (1985). *General Equilibrium Analysis, Studies in Appraisal*, Cambridge University Press.

Weintraub, Sidney (1978). *Keynes, Keynesians and Monetarists*, Philadelphia: University of Pennsylvania Press.

Weiss, L. W. (1980). 'Quantitative studies of industrial organisation', in M. Intrilligator, ed., *Frontiers of Quantitative Economics*, Amsterdam: North Holland.

Weitzman, Martin (1985). 'The simple macroeconomics of profit sharing', *American Economic Review*, 75(5), December: 937–53.

Whitman, R. (1941). 'A note on the concept of degrees of monopoly', *Economic Journal*, 51:261–9.

Wood, Adrian (1975). *A Theory of Profits*, Cambridge University Press.

Wright, E. O. (1977). 'Alternative perspectives in Marxist theory of accumulation and crisis', in J. Schwartz, eds., *The Subtle Anatomy of Capitalism*, Santa Monica, California: Goodyear.

INDEX